T0034351

Between Doom
and Denial

Between Doom and Denial

Facing Facts About Climate Change

SUTHERLAND
HOUSE

TORONTO, 2023

Sutherland House
416 Moore Ave., Suite 205
Toronto, ON M4G 1C9

Copyright © 2023 by Andrew Leach

All rights reserved, including the right to reproduce this book or
portions thereof in any form whatsoever. For information on rights and
permissions or to request a special discount for bulk purchases, please
contact Sutherland House at info@sutherlandhousebooks.com
Sutherland House and logo are registered
trademarks of The Sutherland House Inc.

First edition, October 2023

If you are interested in inviting one of our authors to a live event or
media appearance, please contact sranasinghe@sutherlandhousebooks.com
and visit our website at sutherlandhousebooks.com for more
information about our authors and their schedules.

We acknowledge the support of the Government of Canada.

Manufactured in Canada
Cover designed by Serina Mercier & Jordan Lunn
Book composed by Karl Hunt

Library and Archives Canada Cataloguing in Publication

Title: Between doom & denial : facing facts about
climate change / Andrew Leach.
Other titles: Between doom and denial
Names: Leach, Andrew (Andrew J.), author.
Description: Series statement: The McGill Max Bell lectures
Identifiers: Canadiana (print) 20230458327 | Canadiana (ebook) 20230458343 |
ISBN 9781990823497 (softcover) | ISBN 9781990823503 (EPUB)
Subjects: LCSH: Climate change mitigation—Canada. |
LCSH: Climatic changes.
Classification: LCC TD171.75 .L43 2023 |
DDC 363.738/746—dc23

ISBN 978-1-990823-49-7
eBook 978-1-990823-50-3

THE McGILL MAX BELL LECTURES

Economic Ideas for a Stronger Canada

The newly-created McGill Max Bell Lectures are devoted to examining and discussing policy ideas to strengthen Canadian prosperity. Our country has always faced its fair share of challenges, whether homegrown or from our participation in a complicated global environment. Recognizing these challenges and thinking about them clearly is the first step. The next step is to design and implement the public policies we need to deliver the shared prosperity we seek. These lectures are designed to improve our progress along that path.

The McGill Max Bell Lectures will be held in three Canadian cities annually and the associated book will be published by Sutherland House. These lectures have been made possible by donations to the Max Bell School with a generous lead gift from Mr. Thomas E. Kierans, O.C., a proud McGill graduate from the Faculty of Arts in 1961.

For Will and Caroline. May the world you inherit
be one with less doom and less denial.

Acknowledgements

First, thank you to Chris Ragan and the Max Bell School of Public Policy at McGill University for the opportunity to deliver the inaugural Max Bell Lectures. And thank you to Ken Whyte and the team at Sutherland House for your patience, diligence, and flexibility.

This book would not be what it is today without the incredible research assistance I received from Max Bell School students Vivian Allison and Jimy Beltran. Vivian and Jimy did more than just research: they edited text, workshopped ideas, read countless drafts, and provided much-needed encouragement and positive reinforcement to help get this project to the finish line.

The structure for this book was inspired by a talk given by Aaron Hughes on his book *10 Days That Shaped Modern Canada*. I loved how Hughes' work motivated the audience to think about their own lists of important dates in Canadian history. I hope to motivate my readers to think of their own lists of little lies and half truths that permeate discussions of climate change in Canada.

I've benefited from the work of many authors as I pulled this book together. Chris Turner, Katharine Hayhoe, John Vaillant, Nic Rivers, Jeffrey Simpson, Leah Stokes, Kent Fellows, Aaron Cosbey, Rachel Samson, Dave Sawyer, Martin Olzsynski, George Hoberg, and Kathy Harrison among others have shaped my thinking on climate change and the policies to combat it. Thanks to Jen Winter, Kevin Milligan, Blake Shaffer, Tim Weis, Sara Hastings-Simon, and Trevor Tombe for your encouragement, edits, references, and challenges. And, thanks to Gerry Butts for always pushing me to think about both the climate math and the coal miners.

Finally, to Alison, thank you for making it possible to fit this and all of my other pursuits into our lives. And thank you for your edits, challenges, opinions, and constant encouragement.

Contents

1 | Introduction

Many years ago, on a research trip to Laramie, Wyoming, I spent an afternoon with a colleague at The Library. That The Library is a sports pub and brewery does not make this statement any less accurate, but it might change your impression of the amount of work accomplished that afternoon. Many other university towns have similar establishments. Why? In no small part for the joy of telling your parents that you were at The Library.[1] It is not a lie, at least on its face. It is a half-truth. A lie by omission. An easy way to avoid what would otherwise be a difficult conversation. This book tackles a series of similar half-truths, lies by omission, and too-clever-by-half excuses that we, as Canadians, deploy when talking about climate change.

Climate change is the environmental, political, and societal challenge of our time. It is a challenge to which we, as Canadians, have not always responded with best efforts and grand ideas. Too often, our discussions of climate change devolve into denial or countervailing promises of doom that, while comforting for some, serve little purpose as we prepare for the future. We tend to shrink from the challenge or push back with soundbites about Canada's place in the world. We are a cold country, so the coming changes will not affect us, and we might even benefit from a warmer planet. We contribute less than 2 percent of global emissions, so anything we do will not matter. We can adapt to climate change at a lower cost than mitigating emissions. We tell ourselves, and in some cases our shareholders, that the world will still use oil and gas, as though that provides assurance that our fossil fuel industries will not be affected by action on climate change. Some of us are prepared to dismiss the substantial potential of wind and solar power because we imagine that its usefulness is compromised by sunsets and calm days. And some are quick to imagine that government action can create new industries to replace lost jobs, wages, and tax revenues, and provide a *just transition* away from fossil fuels.

These are, at best, half-truths. Like an evening studying at The Library.

Canada is unlikely to see the same catastrophic level of damage as other countries; in fact, credible estimates do see Canada benefiting from climate change, although I will take issue with some of their conclusions. As Canadians, we have the wealth to buffer ourselves better than most through investments in adaptation. The world will not stop using fossil fuels tomorrow, and overly stringent policies in Canada might make our energy transition more expensive than it needs to be. Renewable energy sources do face challenges when it comes to meeting the energy demands of our cold, northern winters. New energy technology will change our lives and, just as we have altered our lives to match the technology of the twentieth century in ways previously unimagined, we will change our lives again to adapt to the technology of the twenty-first century. And, while promises of a just transition probably offer false hope to those most likely to be affected and allow decision-makers to ignore the real costs of a policy-forced energy transition, our social safety net has and will continue to buffer workers from the worst consequences of economic loss. Our economy will need to evolve in substantial ways; it is facing a transition unlike anything we have seen in the past. I believe that Canada is up for the challenge.

This volume is the first installment of a new annual book and lecture series at the Max Bell School of Public Policy at McGill University in Montreal. It is limited in scope and targeted in its approach. I have been working on climate change policy in Canada since the early 2000s, mixing academic and public writing with consulting and policy advisory roles. I was seconded to the federal (2012–13) and Alberta provincial (2015) governments to work on policy implementation. Over that time, I heard endless explanations, rationalizations, and excuses for why Canada should or should not act on climate change. They all inspire this book. I could give dozens of examples of our failure to meet the challenge or to face facts, but I have chosen to concentrate on six. My list is not exhaustive—I have left off dozens of items—and others may see different challenges in our climate change discourse. That's fine. I have structured the book around what strike me as the most pervasive and/or insidious lies and half-truths that permeate the climate change debate in Canada today, as well as those that will be best suited to delivering an interesting series of lectures for the Max Bell School. My objective is not to have the last word but to spur discussion and challenge all of us to think harder about the coming transition.

* * *

Climate change is an enormous subject. It blends science, economics, international geopolitics, and ethics. The physics of anthropogenic climate change are well-established. As Katharine Hayhoe explains in her book *Saving Us*:

The Earth is wrapped in a natural blanket of heat-trapping gases. Most of the Sun's energy goes right through this blanket, as it does through a window, heating the Earth. The Earth absorbs the Sun's energy. It warms up, and gives off heat energy. The blanket traps the heat energy, keeping the Earth around 33°C or 60°F warmer than it would be otherwise. . . . Hundreds of years' worth of carbon dioxide, methane, and nitrous oxide emissions have artificially increased the thickness of that natural blanket.[2]

Emissions of greenhouse gases (GHGs)—the carbon dioxide, methane, and nitrous oxide that Hayhoe lists—accumulate in the atmosphere and cause the planet to warm. The resulting increases in global temperature disrupt our planet's natural systems. These changes affect us all, regardless of where, when, or for what purpose the emissions occurred. Climate change is the prototypical common property problem. Just as societies of the past have over-fished, over-grazed, or over-hunted and depleted the natural capital on which they depended, we are over-taxing the capacity of our planet's systems to absorb and sequester GHGs.

The scientific evidence of the causes and impacts of climate change is getting more robust and more worrisome. The Sixth Assessment Report (AR6) of the Intergovernmental Panel on Climate Change (IPCC), the leading authoritative summary of the state of global climate science, presents stark conclusions.[3] Global average temperatures are well above historical (1850–1900) values, and their recent rate of increase has been faster than in any other period in the last two millennia.[4] The warming we have seen is mostly human-caused.[5] It is not the sun, volcanoes, sunspots, orbital cycles, or whatever other theory you might have seen on Facebook; it is us.[6]

Anthropogenic (human-caused) increases in global average surface temperatures are estimated to be a little over 1°C to date.[7] And global emissions of the GHGs that cause climate change are increasing (other than a decline in 2020 due to COVID-19), although thankfully at a slower rate compared to the early 2000s.[8] The warming brought about by increased concentrations of GHGs in our atmosphere is causing substantial change to many natural systems on our planet. While the attribution of any one event to climate change remains a challenge, the broad conclusion of the IPCC report is that widespread adverse impacts and losses are already a reality and are affecting more vulnerable populations disproportionately.[9] As Katharine Hayhoe writes, this was all both predictable and predicted:

If you compare climate predictions with what really happened over the past few decades, you will find that the scientific community gets the changes in global temperature right. But studies have found that it tends to underestimate other observed changes and their resulting impacts.[10]

Global average surface temperature changes are used as the primary metric for the severity of global climate change and as the primary way to frame goals for global climate change mitigation. The Paris Agreement, adopted by most of the world, includes the goal of "holding the increase in the global average temperature to well below 2°C above pre-industrial levels and pursuing efforts to limit the temperature increase to 1.5°C above pre-industrial levels."[11]

As the science of climate change has become clearer and as estimates of potential damages have increased, the global community has responded, although not with the urgency required to assure that we meet the goals set out in the Paris Agreement.[12] National commitments to reduce emissions have become more ambitious, and low-emission energy technologies are getting cheaper faster than almost anyone expected. Because of these changes, what we might previously have imagined as a worst-case scenario for climate change is now viewed by many as an unrealistic exaggeration.[13] Meeting global climate goals, which once seemed merely aspirational, is now plausible, but only if we implement substantially more stringent policies and sustain them over decades.[14]

Aggressive global action faces substantial challenges. Climate change mitigation is the purest of public goods: it will benefit every country and region, and our enjoyment of a more stable global climate will not reduce the capacity for others to enjoy those same benefits.[15] This makes the problem difficult to solve: there is limited incentive for any individual nation to mitigate its emissions since it can enjoy the benefits of others' mitigation efforts. The economic incentives tell us to free ride and wait for others to act. Unfortunately, the longer the world waits to act aggressively on climate change, the greater the locked-in damages.

Action on climate change has already fundamentally changed our global energy systems. There is no business-as-usual in energy. Stunning growth in renewable electricity generation and the widespread adoption of electric vehicles are now the baseline. We have moved from talking about peak oil supply, in which physical and economic constraints restrict the amount of oil production, to peak oil demand, where consumer alternatives and policy preferences shape the future of the oil market. Whereas forecasts for global oil consumption not long ago topped 120 or even 130 million barrels per day, mainstream forecasts now see little to no growth beyond today's level of around 100 million barrels per day, even without more substantial action on climate change.[16] Still, this is nowhere near enough progress to meet global climate goals.

We face a substantial, sometimes mind-numbing gap between current global actions to mitigate climate change and the sustained efforts required to meet global emissions reduction goals. According to the latest IPCC report, the measures we have undertaken to date would still see global emissions increase steadily, though more slowly, through 2050. National commitments to emissions reductions through 2030 announced at global climate conferences in Paris and Glasgow, if fully accomplished, would only be sufficient to limit global temperature increases to 2.8°C by 2100.[17] The path to keeping global average temperature increases below 2°C by 2100 requires cutting emissions at increasing rates between now and 2050. Assuming we reverse emissions growth and start seeing small decreases of roughly 1 percent per year between now and 2030, we would then need to cut emissions by close to 4 percent per year through 2040, accelerating to more than 6 percent per year from 2040 through 2050 and beyond.[18] If we want to keep temperature changes below 1.5°C, the path is substantially more challenging. Some countries have made ambitious long-term pledges which, when combined, could be sufficient to limit global temperature increases to less than 2°C, but these have yet to be transformed into durable policy solutions. And the longer we delay the implementation of the necessary policies, the harder meeting these targets becomes and the more we are left to choose between faster future emissions reductions or even more serious future climate change damages.

* * *

Canada faces many of the same barriers to climate action faced by the global community. In fact, Canada provides a microcosm for many of the world's climate change challenges.

We are a small part of the global commons and account for a small share of global emissions, so the actions we take as a country may seem insignificant. This sense of irrelevance mirrors that of individual Canadians, provinces, and municipalities in the face of our daunting national emissions reduction goals. Different parts of the country face different risks from climate change, have access to different means to reduce emissions, and have different historical emissions trajectories, just as countries around the world differ along these same lines. As a result, Canada's different parts act similarly to members of the global community, arguing over the level of responsibility that each should bear in reducing emissions. Our provinces and cities have often set targets for future emissions reductions without the policy ambition to deliver on those goals, again like Canada and many other countries.

Canada has increased action and ambition on climate change over the past few years. National carbon pricing, clean fuels and clean electricity regulations, vehicle emissions standards, enhanced climate change risk disclosure rules, and myriad subsidies for emissions-reducing capital investment are bending our emissions trajectory downward. But aggressive action on climate change continues to be opposed by some provinces, in particular Alberta and Saskatchewan, where emissions per capita are dramatically higher than anywhere else in the country and where provincial economies rely more on fossil fuels. As is the case in most nations, some of Canada's policy progress could be reversed with the next election, although the implementation of many now-cheaper low-emissions solutions like renewable power and electric vehicles is unlikely to backtrack substantially. A majority of Canadian voters agree on the need to act on climate change; opinion is more varied as to how we should act and how aggressive our measures should be.

Despite substantial technological and policy progress, Canada remains far from meeting its targets. It is a familiar refrain: we commit to aggressive targets at international meetings but lack the policy ambition at home to create the conditions to meet those targets. At the United Nations climate change conference in Glasgow in 2021, Canada committed to a 40–45 percent reduction in emissions relative to 2005 levels by 2030 and reiterated its goal to achieve a net-zero emissions economy by 2050.[19] As shown in Figure 1, projections from Environment and Climate Change Canada hold that our current and planned policies (labelled "2022 ECCC Reference Case") would only produce a 14 percent emissions reduction below 2005 levels by 2030, or about one-third of our Glasgow commitment. More ambitious projections (labeled "ECCC Additional Measures Case") see a 30 percent drop in emissions below 2005 levels by 2030, but assume that substantial new policies are implemented. Neither projection sees Canada meet its 2030 targets. And if we are not on track to meet our 2030 targets, our long-term goal of net-zero emissions by 2050 becomes much less attainable.

The headline, then, is that Canada is undershooting yet another international target, but all may not be lost. If the projections in the 2022 Additional Measures Case come to fruition, we could meet the target to which we committed for 2030 at the Paris climate conference in 2015. It's not as ambitious as the Glasgow (2021) target, but it is progress compared to our efforts to meet our previous Kyoto (1997) and Copenhagen (2009) targets. The latter targets were set by the Chretien and Harper governments, respectively. Each failed to implement the policies necessary to deliver on their promises. With more ambitious policy actions and the rapid deployment of clean energy, we may finally begin to meet our commitments.

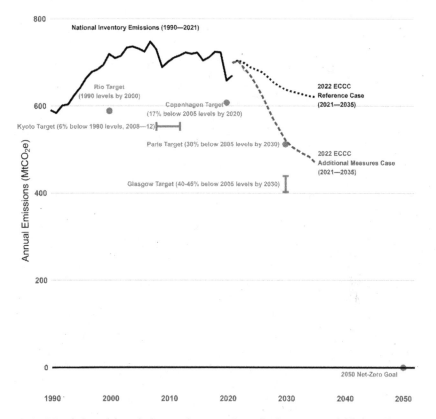

FIGURE 1 | Canada's emissions and targets. Data: Environment and Climate Change
Canada. Author's calculations and graph.

But we have a long way to go and many crucial issues to address. We are far
from consensus on such questions as how large or how rapid an energy transition we need to undertake, whether colder countries can avoid the costliest
consequences of climate change, or why small countries with low emissions
need to bother reducing them at all. Real debate remains on the future role of
intermittent sources of clean energy like wind and solar power, just there is
uncertainty over future demand for our oil and gas even as some demand that
we leave it all in the ground. And we do not know that we can navigate the
coming energy transition without leaving anyone behind. These questions are
the subject of this book.

2 | Climate change will not be that bad, and we can adapt

Climate change mitigation will be costly. But as John Holdren, director of the White House Office of Science and Technology Policy, said often during the Obama years, mitigation is only one of our three choices for dealing with climate change.[20] The other two are adaptation and suffering. According to Holdren, "the more mitigation we do, the less adaptation will be required, and the less suffering there will be."[21] There are others, however, who forget this zero-sum equation and instead leverage the potential for adaptation to argue for less, or even no, efforts to reduce emissions to mitigate climate change.

> It is easy to construct climate disasters. You just find a current, disconcerting trend and project it into the future, while ignoring everything humanity could do to adapt.
> Yes, the climate is likely to change, but so is human behavior in response.[22]

There is truth in this quote, from a column in the *Wall Street Journal* by self-described climate skeptic Bjorn Lomborg. If there were not, it would be easy to disregard. But certain omissions make his quote insidious.

It is obvious, as Lomborg states, that any study will over-estimate potential damages if it assumes that adaptation does not occur in response to climate change. I agree that we should strive for better estimates of the costs of climate change and make sure that these estimates take account of our capacity for adaptation. Those estimates should guide our policies and strategies.

But is that the point Lomborg is looking to make? Is he rightfully arguing for better damage estimates? Or is he discouraging mitigation and trying to convince you that, perhaps, climate change will not be as bad as you may have heard?

Adaptation can reduce the costs of climate change for some, but it will not eliminate them. And the ability of wealthier populations to invest in adaptation will likely further skew the distribution of the costs of climate change toward the world's poorest countries. There are substantial benefits to be had from both adaptation to and mitigation of climate change. Early investments in adaptation prepare us for the impacts of climate change that we have already locked in, while mitigation acts to reduce future climate change and future costs. Both can be cost-effective, although the benefits of adaptation can be tailored to benefit those making the investments. Perhaps that is the point of Lomborg's quote: to tell his readers in the *Wall Street Journal* not to worry; they will be able to adapt.

Again, climate change mitigation is a global public good.[23] Each tonne of emissions imposes global costs, and each tonne avoided confers global benefits. The costs of climate change mitigation, however, are local. Governments, businesses, or individual consumers bear the costs of investments to reduce emissions and cannot appropriate the full, global benefits of their actions. The costs *and* benefits of adaptation, on the other hand, are both more local, which is part of the attraction. Spending on adaptation is like buying yourself climate change insurance.

However, it is not a panacea.

The benefits of our individual and collective actions to reduce emissions are much broader than the benefits of adaptation because they are global rather than local. This is the upshot of the public good nature of mitigation: everyone benefits. While adaptation investments might secure one municipality against the worst effects of sea level rise, mitigation efforts reduce the severity of sea level rise for everyone. And they also reduce all the other impacts of climate change as well.

The mitigation efforts already underway, while impressive, are not enough. They are potentially sufficient to stabilize global emissions at roughly current levels.[24] And the combination of Nationally Determined Contributions as updated through the Paris and Glasgow climate meetings and other long-term pledges may be sufficient to keep long-term temperature changes below 2°C, if countries follow through on their commitments.[25] These efforts will spare us from the worst-case scenarios of climate change, but our emissions trajectory will still subject us to a substantially changed climate. The latest IPCC report suggests that we are likely to see between 2.2°C and 3.5°C of warming if long-term commitments to reduce emissions are not met, although there remains substantial uncertainty associated with this range.[26] That much warming promises us a radically different world. A world with a whole lot more of that third option that John Holdren warned us about: suffering.

Human health impacts from high heat and humidity will increase dramatically in a much warmer world. Equatorial regions will be most affected, and increasing swaths of North America will also become dangerously hot and humid if we do not do more to mitigate emissions.[27] There are other impacts, too, that are not as amenable to adaptation as increased heat and humidity. More warming increases the frequency and intensity of extreme weather: heatwaves, heavy rainfalls, and droughts become more likely in certain regions. The intensity of tropical cyclones will grow. Sea ice and glaciers will melt, and the Arctic permafrost will thaw.[28] And wildfire smoke will increasingly fill our summer air. There are limits to our capacity to adapt to these challenges.

Our food supply is also vulnerable to climate change. Adaptation strategies such as the substitution of crops will help in some cases, as might more irrigation, but yields of some of the world's most important staples are expected to drop substantially with worsening climate change, increasing food prices, and worsening food insecurity. And while we can change cropping choices on land, we cannot relocate fish stocks in the oceans. Material drops in fisheries yields are expected to result from climate change, with impacts increasing dramatically with greater degrees of warming. The worst of each of these effects will only be avoided through mitigation. The alternative, for many, will be more suffering. The impact of these changes will be catastrophic in some of the world's poorest regions, but no region of the globe will be spared.

Each of these effects speaks to the insidious nature of calls for adaptation rather than mitigation. While adaptation allows some of us to deal with some of the effects of climate change, only much more investment in mitigation will spare us all from most of its costs.

* * *

Economic evidence argues for much more mitigation than we are undertaking today, not for a strategy based largely on adaptation. Canada and the rest of the world are underinvesting in mitigation because of what economists term a market failure. Emissions of GHGs are costly, and estimates of those costs have increased over time.[29] But the costs of emissions are largely external to the day-to-day and long-term decisions made by consumers and firms. While consumers and firms do not pay the costs of emissions directly or indirectly, that doesn't mean those costs go away. The costs will still be felt through future climate change damages, and we are incurring more of those future costs than we should. The costs of climate change that we will collectively incur will be greater in most cases than the costs we could collectively incur now to avoid the emissions altogether. The challenge lies in motivating collective action to yield collective gains.

The economics of climate change or most any other form of pollution are relatively simple. When private costs—the costs that consumers or firms pay directly or indirectly for goods and services—differ from the total costs that producing those goods and services imposes on society, including the costs of pollution, markets no longer work efficiently to determine how much should be produced and consumed. Where no one is liable for the costs of pollution, there will be too much of it.[30] And where polluters are not held to account for the damage caused by their activities, innovations that reduce emissions have less value than they should; if companies cannot profit from reducing pollution, they will not be willing to pay for the technology to do so. Government policy, whether in the form of regulations, taxes, or subsidies, can correct for these market failures and lead to better and more economically efficient decisions by rewarding mitigation and ensuring that emissions have costs that reflect their true impact on society. If consumers and firms were liable for the impacts of their actions on the environment, they could then properly weigh the benefits of mitigation versus adaptation. In the absence of such policies, mitigation is always going to be undervalued. Today, firms and consumers are not empowered by the market and, generally, not even by policy to make economically efficient choices.

When economists consider the value of climate change mitigation, they rely on the concept of a social cost of carbon. The social cost of carbon is an estimate of the future costs of incremental emissions of GHGs, translated into present-day dollars. Think of it as a measure of the amount we, collectively, should be willing to pay to avoid those emissions, or the amount we should charge ourselves for the privilege of releasing them. We are, in most cases, not even close to paying or charging enough.

Estimates of the social costs of GHG emissions are shockingly high and getting higher every year.[31] A new assessment by the US Environmental Protection Agency, subsequently adopted for use in assessing the value of Canadian efforts to reduce emissions, estimates that each tonne of carbon dioxide emissions causes $261 in damages, a dramatic increase from previous estimates.[32] What these numbers tell us is that, collectively, we should be willing to pay up to $261 per tonne—roughly $0.60 for the emissions released when burning a single litre of gasoline—to avoid incremental GHG emissions. The numbers put a dollar value on the suffering to come, as they measure damages from emissions that we do not eliminate through mitigation. And yes, they account for our capacity to adapt.

These updated estimates of the social cost of carbon, while shockingly high, are almost surely an underestimate. First, consider what is included. Temperature mortality accounts for roughly half of the estimated social costs of climate change.[33] In Canada, we are relatively wealthy and our climate is cooler than most, which makes it inviting to imagine that adaptation is simply a matter of turning up the air conditioning or turning down the heat. For much of the world, that is not an easy option. Most of the rest of the estimated damages is due to agricultural impacts around the world.[34] Again, in Canada, warmer temperatures will bring longer growing seasons, but in much of the world, warmer temperatures will be catastrophic for agricultural yields. Sea level rise, another source of damage explicitly modeled in the estimates, has a relatively small impact in part because it occurs far into the future and is more amendable to adaptation relative to other sources of climate damage. Similarly, impacts on heating and cooling costs are relatively small due to the benefits of lower heating costs in some regions offsetting higher cooling costs in others.[35] As you might suspect, these are far from the only potential costs of climate change.

Estimates of the social costs of carbon estimates are alarming in and of themselves, but they only account for part of the suffering that awaits us as a result of climate change. The effects not accounted for in the updated estimates should give us all further pause. The estimates do not account fully for the costs of many highly disruptive effects of climate change, such as increased extreme weather, changes in precipitation, or increases in wildfires.[36] They do not account for changes in labour productivity, pressures from migration, or geopolitical upheaval. They do not put a value on lost biodiversity or ecosystem services, including losses due to ocean acidification. They do not account for the potential impact of more destructive coastal storms. They do not account for changes in the availability of fresh water, changes in precipitation, or drought patterns. The list goes on. Of course, the estimates are subject to substantial uncertainty, but the social cost of carbon and the long list of omitted effects leave no doubt about the substantial economic benefit of aggressive actions to mitigate climate change.

* * *

There are too many merchants of doom among those pushing for greater action on climate change, and too many exaggerations that are accepted because they push us in the right direction, toward greater action on climate change. My first piece of public writing to get any mainstream exposure took issue with hyperbolic claims made by former NASA scientist James Hansen and environmental leader Bill McKibben tying the Keystone XL pipeline project to the potential emissions from extracting the entire oil sands resource.[37] I concluded my long piece by arguing that, while every action counts in the fight against climate change, blatant exaggerations are "the arguments most likely to be ignored as alarmist."[38] I have some sympathy for Lomborg's crusade against doomism.

But there is a point where skepticism evolves into something more dangerous. In advocating for adaptation over mitigation, are those who follow Lomborg truly being pushed toward more cost-effective responses to climate change, as he states is his aim? Or are people being pushed away from measures to combat the problem, convinced that it will not be that bad because we can adapt?

Lomborg writes that "adaptation is much more effective than climate regulations at staving off flood risks" and offers a comparison in which targeted spending on adaptation would reduce flooding displacement substantially more than "stringent regulations that keep the global temperature rise below 2°C."[39] Strictly speaking, his statement might be accurate. But it ignores the fact that every tonne of emissions reduced through those regulations would reduce all other climate damages in addition to the flood mitigation benefits, and that all of these benefits add together to provide a meaningful comparison to the costs of mitigation. That calculation is missing from Lomborg's piece. Lomborg's writing also suggests, although it does not explicitly state, that flood control measures and broader climate mitigation cannot be combined. That we have the choice of one or the other. The argument sets up a false dichotomy.

We will get more value from our adaptation dollars with more action to mitigate climate change, and many cost-effective actions to mitigate climate change are available today. In fact, many would even be cost-saving. For example, the International Energy Agency (IEA) estimates that over half of the global emissions of methane (a much more potent GHG than carbon dioxide) from fossil fuel production could be eliminated without incurring any net costs at all, once the value of the captured methane is taken into account.[40] And, we continue to subsidize emissions by spending hundreds of billions or even trillions of dollars per year globally subsidizing fossil fuel consumption.[41] As technologies such as renewable electricity, electric vehicles, and heat pumps improve, the list of cost-effective mitigation opportunities will get longer each year. We should not pass these by because someone tells you that you can avoid the effects of climate change by building a few dikes, moving further inland, or turning up the air conditioning.

Mitigation does more than just reduce our need to adapt. It gives us a fighting chance. As Katharine Hayhoe writes in Saving Us, "For our modern world, the difference between a higher versus a lower emissions future is nothing less than the survival of our civilization."[42]

3 | Canada is a cold country

Finally, someone proclaimed an obvious truth that few dare to utter publicly. According to Moody's Analytics, Canada will benefit from climate
change. Although it will shock many, this forecast should surprise no one.
Canada is a very large, cold country, with . . . an enormous agricultural
potential if the land warms up. There will also be new opportunities for
oil, gas and mineral development in the Arctic. And let us not ignore the
greater personal comfort of living in a more hospitable climate.[43]

That quote is from a 2019 op-ed by Canada's former Finance and Natural
Resources Minister Joe Oliver, trumpeting a study by Moody's Analytics,
which found that Canada was in a position to benefit from climate change.[44]
The Moody's study Oliver cites even finds that the more severe global climate
change turns out to be, the more our gross domestic product will increase.
And, perhaps shockingly, these conclusions are no outlier.

Academic studies published in top journals reach the same conclusions
as the Moody's study that Joe Oliver cites. In 2018, Katharine Ricke and her
co-authors concluded that Canada would benefit from climate change.[45]
Similarly, Lee Hannah and co-authors found that a warming world will provide
"opportunities for economic development that, if done properly, may reduce
poverty and food insecurity in some economically marginal parts of the world,
such as Northern Canada."[46] Another study by Roson and Sartori (which
informed the Moody's analysis) says that Canada will benefit from climate
change primarily due to increased agricultural productivity and tourism.[47]
These are far from the only studies to reach such conclusions.[48] Joe Oliver is
far from alone in asserting that Canada could benefit from climate change.

You are likely thinking that there must be something missing here. And
you are correct.

First, there is disagreement in the literature with respect to what costs
Canada faces from climate change. For example, a recent study by Matthew
Kahn and co-authors predicts that Canada will see 2–13 percent declines in

GDP per capita by 2100 under moderate to severe climate change.[49] Estimates that Canada will benefit from climate change are not outliers, to be sure, but neither are they the universal conclusion of careful study.

Second, and more importantly, studies of the sort I've cited above do not give a complete assessment of how climate change will affect Canada because of the approach used to compute their estimates. The first step in this type of study is to determine country-specific temperature and precipitation responses to GHG emissions for a range of potential global emissions trajectories (how much warmer and how much drier or wetter Canada is likely to be). The next step is to translate predicted climate outcomes into predicted changes in economic output (if it gets this much warmer and this much drier or wetter, this is what happens to economic output).[50] These relationships are generally estimated using data on how countries have fared in years when temperature and precipitation deviated from historical averages. It is here that northern latitude countries (e.g., Canada or Russia) have sometimes been estimated to fare relatively well in a changing climate: they have vast land mass with a lot of agricultural production, so warmer years with longer growing seasons are generally better.[51] It is also here that these studies are likely erring in their estimates of the future costs of climate change.

Statistically, these studies are asking a subtly different question than whether Canada will benefit from climate change. Since they are using data on historical temperature deviations from average conditions, they are evaluating whether Canada has historically been better off in warmer-than-average years and using that to approximate the impacts of a long period of sustained climate change. These are not comparable. Sustained climate change brings with it myriad costly consequences that we would not expect, or not expect to the same degree, in a single warmer-than-average year.

Canada has already experienced a 1.9°C increase in average temperatures from 1948 through 2021, and nine of Canada's ten warmest years have occurred in the last decade.[52] In line with expectations of climate scientists, temperatures have increased in Canada at more than twice the global average,[53] with northern regions of the country warming three times faster.[54] Regardless of future global actions to reduce GHG emissions, these climate changes are going to continue for decades to come. And these sustained temperature changes have had and will continue to have profound impacts across the country—effects that would not be observed in a single warmer-than-average year—including more wildfires and wildfire smoke, the expansion of pests and diseases, changing precipitation and wind patterns, the loss of permafrost and sea ice, and rising sea levels. When all this is factored in, it seems to me very unlikely that Canada will experience anything like a benefit from climate change.

Let's look a little deeper at just some of the effects we should expect from sustained climate change in Canada.

* * *

Permafrost is ground that remains frozen for at least two years.[55] More than 50 percent of Canada's land mass contains permafrost.[56] Northern Canada is literally built on permafrost. Homes, businesses, roads, airport runways, and utility infrastructure rely on it as we might rely on concrete and steel. In many regions, permafrost is no longer expected to remain a solid base of support in a warming climate. Permafrost thaw is already affecting Canada's North. Damaged roads are making it more difficult to travel, limiting access to food, essential supplies, and medical services. Permafrost thaw is impacting homes and buildings in northern communities, increasing costs of construction, maintenance, and insurance. And where permafrost thaw leaves unstable, swampy ground, it limits access to activities essential for food security and for the cultures of Indigenous communities.

We have only just seen the beginning of the permafrost thaw that awaits us. More disruptions to transportation and building infrastructure and changes to the land will occur as the climate warms. They will be irreversible in our lifetimes. Costly interventions may be able to mitigate some of the damage, but the economic and cultural costs of the thawing permafrost will be substantial. Because permafrost loss would not be significant in a single warmer-than-average year, it is not considered in the macroeconomic studies used to forecast the likely impacts of climate change in Canada.

Like permafrost, annual freeze–thaw cycles are very important to life in Canada, and these cycles will be fundamentally altered by climate change. For example, climate change is already affecting users of winter roads in the North and near-North.

Winter roads are essential lifelines into northern communities, bringing food, fuel, building materials, and other essential goods in while allowing the export of traded goods and a relatively low-cost route to larger centers in the South for community members. Warmer temperatures could "dramatically decrease the duration and extent of winter ice road coverage, by as much as 99%–100%."[57] Even moderate climate change is expected to have a dramatic effect on the open seasons on winter roads, which, in some cases, are already very short.[58] As climate change worsens, these roads will have to be replaced with year-round roads, or the communities they serve will be forced to rely on much more costly air transfers from the South. Like the thawing of permafrost, the costs of long-term changes in freeze–thaw cycles would not be accurately estimated by analysis focused on impacts arising in individual warmer-than-average years.

Climate change will not only affect infrastructure serving remote communities in the North. The resource industry, from mining to oil and gas extraction, relies on frozen ground to work in regions that would otherwise be inaccessible. Permafrost also serves to secure tailings ponds and other mine wastes that may present risks as temperatures warm.[59] Again, effects like these are likely to be missed in studies linking short-term temperature fluctuations to economic activity.

In addition to frozen land, sea ice is critical to the way of life in many northern communities. As the climate warms, Artic sea ice is receding at a rate of more than 12 percent per decade. More of the total ice cover each year is seasonal ice rather than long-term ice.[60] That means, as shown in Figure 2, the average age of sea ice is getting younger. *Younger* seasonal ice means the ice cover responds more to weather, making coastal ice patterns more volatile. This has food security implications: there is less dependable ice cover for traditional hunting and fishing activities.

Scientists predict we are likely to experience the first ice-free Arctic summer before 2050.[61] Sea ice loss has serious consequences that could extend beyond the loss of traditional ways of northern life. It could set off important feedback loops: warmer seas release more methane, a potent GHG, while water reflects less sunlight than ice, allowing more of the sun's heat to be absorbed in the ocean water. Each of these impacts leads to a further acceleration of warming. Talk about a vicious cycle.

* * *

Studies that show Canada benefiting from a warmer world assume many things remain as they are while we benefit from longer growing seasons. In fact, the oceans that abut Canada from the North, the East, and the West are expected to change significantly as the climate warms. And, as with thawing permafrost, the impacts could be catastrophic.

Increasing global average temperatures and the melting of polar ice caps and mountain glaciers will have an impact on sea levels. Sea levels today are already roughly 25 centimetres above 1900 levels, and, depending on climate change mitigation actions, they could rise another 25–75 centimetres by 2100.[63] Sea level rise of at least 50 centimetres by the end of the century is relatively certain, and further increases will continue far into the future. Even with aggressive action on climate change, a cumulative sea level rise of 2 to 3 metres over the next two millennia is possible, although the latest IPCC report notes low confidence in these long-term predictions.[64]

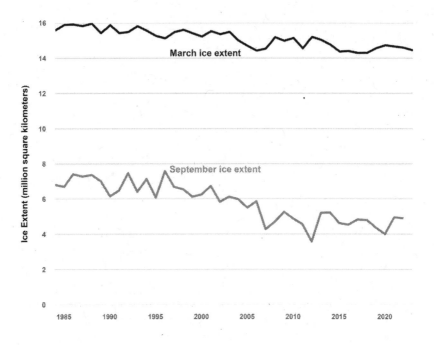

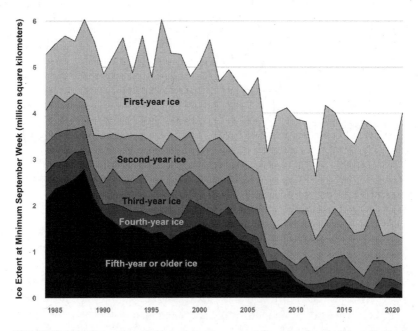

FIGURE 2 | Arctic sea ice winter and summer extent (top panel) and vintage (bottom panel). Source: US Environmental Protection Agency data, author's graphics.[62]

Uncertainties notwithstanding, the impacts of sea level rise will be felt on all three Canadian coasts and will worsen over time and with the severity of climate change. The magnitude of sea level rise expected this century is substantial, and, while we can adapt to rising sea levels, the costs of doing so will be large. More than 330,000 Canadians currently live on land below the high-tide line. That figure could increase to as many as 410,000–840,000 people by 2100 under worst-case climate change scenarios.[65]

Canadian coastal municipalities are rapidly adopting new planning guidelines to deal with predicted sea level rise. The British Columbia government now recommends planning based on 1 metre of sea level rise by 2100 and 2 metres by 2200.[66] Considering, for example, that roughly half the City of Delta (population 111,000) lies less than 1.5 metres above sea level, you see how this becomes a matter of serious concern.[67]

In response, the City of Delta considered three options: more capital-intensive "hold the line" and "reinforce and reclaim" strategies to keep the water out; "managed retreat", allowing the seas to inundate low-lying areas, and the building up of critical infrastructure above flood levels.[68] This example captures the difference between sustained climate change and a warmer-than-average year: the estimates in Joe Oliver's favourite study certainly do not contemplate the costs of raising hospitals, schools, and other infrastructure above new high-tide levels because that's not something we do in response to a single warmer-than-average year.

For communities built at low elevations, a rise in sea level combined with the potential for climate change to lead to more intense extra-tropical storms creates a threat multiplier.[69] In Atlantic Canada, these fears were amplified by the recent passage of Hurricane Fiona in September 2022. Fiona was the costliest storm event in Atlantic Canadian history.[70] When Fiona crashed ashore that fall, the temperature in the North Atlantic was the warmest it has ever been in September. As I write in summer 2023, the North Atlantic is at the warmest temperature ever recorded at this time of year.[71] The exception has become the norm.

Sea level rise will make life very different in some parts of Canada but would generally be ignored by cost estimates based only on extrapolations from how Canada has fared in relatively warm historical years. Sea level rise, like the thawing of permafrost, occurs only as a result of decades of sustained warmer temperatures.

* * *

Just as our oceans and our northern land will change, our forests face a radically different future as the climate changes. A future with more fire and smoke.

The 2023 wildfire season in Western Canada began with a rash of fires in late April and early May. Tens of thousands of people evacuated their homes. It was an eerie reminder of the wildfires that struck Lytton in 2021, Fort McMurray in 2016, and Slave Lake in 2011. In a changed climate, the *fire weather* that preceded each of these events will become more common.[72]

The impact of climate change on the likelihood of wildfires is already being felt in Canada. Megan Kirchmeier-Young and co-authors estimate that wildfire risk in Western Canada was 1.5 to six times higher than would have been the case without climate change at the time of the Fort McMurray fires in 2016.[73] As John Vaillant wrote in the *Globe and Mail*, "In Fort McMurray on May 3, 2016, the temperature broke the record high for that date by 6 degrees Celsius, and the relative humidity plummeted to 12 percent—drier than Death Valley. For fire, that is as good as gasoline."[74] Recent analysis estimates that 37 percent of the cumulative area burned by forest fires from 1986 to 2021 in western North America was a result of human-caused climate change.[75] Under some climate change scenarios, Western Canada's northern boreal forest could see an additional increase of more than 15 days per year of the same hot, dry *fire weather* that supercharged the Fort McMurray fire.[76]

While no individual event can be said with certainty to have been caused by climate change, wildfires have become much more likely and more dangerous than would otherwise be the case. In the words of James Hansen, we've loaded the dice.[77] Future years could well make 2023 look like a mild fire season.

Climate change does not just lead to more *fire weather*. The cumulative impacts of sustained climate change include many other threats to Canadian forests, which, in turn, amplify fire risks. Just as a longer growing season increases agricultural productivity in countries at northern latitudes, it also makes a more hospitable environment for pests.[78] Anyone driving through Western Canada can see evidence of the mountain pine beetle infestations in British Columbia and Alberta.[79] These climate-change-fueled infestations leave behind dead trees, which provide wildfire fuel, amplifying the already-elevated threat from *fire weather*.

And as much of the continent is learning this summer, it is not just fire we should be worried about but smoke too. Asked about the impacts of climate change in 2012, Michael Flannigan, my former University of Alberta colleague and a wildfire expert, concluded that "the future is smoky."[80] In early May 2023, a presentation to the City of Calgary's Emergency Management Committee contained a striking set of statistics that validate Flannigan's predictions: "Between 1953 and 2014, the city never saw more than 100 smoke hours in a year, and the average was 12 hours per year. Since 2015, there have been six years when Calgary exceeded the 100-smoke-hour threshold, and three years where it exceeded 300."[81] As I write this, smoke from the 2023 wildfires is settling in over both Edmonton and Calgary for the first time this year, and, like many Canadians, I am left to wonder how our future summers will look, and what that will mean for our health.

Wildfire smoke, in turn, amplifies impacts and exacerbates drivers of climate change. Increased smoke reduces air quality, as we have all learned in spades in 2023.[82] The reduction in air quality, in turn, increases the use of air conditioners, as open windows become a health hazard.[83] The particulate matter in smoke can trigger asthma attacks, and long-term exposure has been linked to myriad health consequences.[84] To make matters worse, where electricity is provided by fossil fuels, increased air conditioner use leads to increased GHG emissions and other pollution, leading to negative local health effects and exacerbating climate change.

Wildfire smoke can also have a substantial impact on solar electricity generation, and thus can frustrate efforts to mitigate climate change. I tested this with my own household solar power system and found that smoke can cause a drop in solar generation roughly equivalent to the 70 percent solar eclipse that Edmonton experienced in 2017. Similar effects were observed in 2023 as wildfire smoke blanketed the Eastern United States.[85] As with increased air conditioner use, decreased solar power generation due to wildfire smoke can increase the very emissions that lead to climate change, in turn leading to more *fire weather* and more smoke.

Increased fire and smoke, like sea level rise and permafrost thaw, are effects that will not be captured in the individual warm years analyzed in studies of the type that Joe Oliver cites. They are triggered through sustained warming over time. And so the costs of a future with more fire and smoke are not accounted for in many of the studies that claim that Canada will benefit from climate change.

* * *

Heatwaves (or *heat domes*) and extreme rainfall events are also expected to become more likely and more frequent with climate change. As the IPCC reports, it is virtually certain that hot extremes have become more frequent and more intense across most land regions of the world, and human-caused climate change is the main driver of these changes.[86] Studies looking at recent heat waves lay the blame squarely on GHG emissions and the climate changes they induce. Martha Vogel and co-authors conclude that the 2018 Northern Hemisphere heat wave "could not have occurred without human-induced climate change."[87] In the wake of the 2019 European heatwave, similar analysis concluded that "the likelihood of such a heatwave in June has increased 100-fold since around 1900, owing to the combined influence of climate change and other factors, such as air pollution."[88] An event like 2021 heat dome that affected British Columbia remains extremely unlikely in the current climate but "would have been at least 150 times rarer without human-induced climate change."[89] The extreme heat that gripped Western Canada in June 2021 and again in May 2023 is a further indicator of what is to come. What we have seen to date is only the beginning.

Yet another product of changes in natural systems due to sustained warming is flooding, and, like the effects mentioned earlier, it, too, might not be captured in data for a single year. Nearly 80 percent of Canadian cities are built on floodplains, and urban flooding is expected to have a more substantial socioeconomic impact in a climate-changed world.[90] Worse yet, "only 6 percent of people in Canada who live in flood risk areas are aware that this is the case" because Canada lacks comprehensive mapping of floodplains.[91] And, even if we had flood-risk maps, they would now likely be inaccurate. As former Minister of Environment and Climate Change Catherine McKenna said during recent flooding in Ontario and Quebec, "what we thought were one-in-100-year floods are now happening every five years (or) in this case, every two years."[92] We are living in a different world.

Flooding risk will be exacerbated by climate change almost everywhere in Canada, as both higher total yearly rainfall and individual extreme rainfall events become more likely.[93] That said, it is more challenging to attribute any specific flooding or rainfall event to climate change than it is for a wildfire or heatwave. For example, Bernardo Teufel and co-authors found only limited evidence that climate change would make an event like the 2013 Southern Alberta floods more likely.[94] Similarly, Kit Szeto and co-authors found limited evidence of a link between climate change and the 2014 flooding on the Canadian prairies.[95] Even as the climate is changing, we should use caution when it comes to drawing causal links between specific extreme events and climate change.

Measuring damages from flooding and other extreme weather events presents challenges as well. The economic impact of extreme flooding and other weather events in Canada each year will be small relative to the total size of the economy, but it will be very concentrated in affected regions. It won't feel small to them. Interestingly, the cleanup from flooding and other extreme weather events will often show up as a positive economic impact, which can skew the results of an assessment of climate change.[96]

How can there be positive economic impacts from a flood?

Natural disasters like a flood cause millions or even billions of dollars in insured and uninsured losses. Insured losses, at least in the near term, are dollars that would otherwise have appeared as insurance company profits, to be spent or saved by their owners or equity partners. Uninsured losses are funds that would otherwise be spent by Canadians on other things, perhaps travel, or these losses might force people to work more than they otherwise would, skewing the labour-leisure trade-off in a particular region. These effects can appear positive in GDP terms, an example of what economists call the *broken windows* fallacy. Fixing all the metaphorical *broken windows* in a town leads to an increase in economic activity, and our economic accounts do not differentiate between *good* and *bad* activity. The cleanup from a storm looks a lot like what we might normally call *stimulus* spending in a recession. And, like so much else in this volume, there is some truth behind the lie. Fixing a broken window registers just like buying a new bicycle in GDP terms, as both are linked to new production of goods and services. But, as anyone who has had to replace a broken window knows, doing so does not really make you better off, and the money could have been better spent (e.g., on a new bike). Climate change will create a lot of metaphorical broken windows for Canadians to fix, and while dealing with them will lead to economic activity, it is not the type of economic activity we should welcome.

This is another caveat to keep in mind when considering studies of the economic benefits and costs of climate change. Where studies focus on economic activity, rather than on well-being, they may confound cleanup costs and even expenses incurred to adapt to climate change as economic benefits. In both cases, we would be better off if we did not have to spend resources on such things.

* * *

It should be clear by now that studies based on individual warmer years offer a poor guide to how Canada will fare over the long term with a changing climate. And we haven't begun to look at how we are impacted by climate effects occurring outside our borders, something else largely ignored by these studies. The econometric approach used to estimate climate change costs in Canada implicitly assumes that Canada exists in isolation. Statistically, many of these studies ask whether we would prefer a warmer Canada in a world experiencing average climatic conditions, not a warmer and more volatile Canada in a warmer and more volatile world. They implicitly assume that we can somehow duck the collateral damage from the physical, economic, and geopolitical consequences of climate change.[97]

Canada is a small, open economy with myriad ties to the rest of the world. Climate change is consistently estimated to be costly to the world in general and to the United States, our largest trading partner, in particular.[98] The studies cited in columns like Joe Oliver's do not consider the impacts on Canada of economic losses in the United States, even though we know from experience that American economic upswings and downturns reverberate in our economy regardless of the cause. If the United States suffers economically from climate change, so will we. Remember those social costs mentioned earlier? They are being paid for by someone, and they are captured in studies that show substantial global costs from climate change. Just as climate change will cause economic shocks in the United States that will reverberate into Canada, we will not be immune to the global costs of climate change.

And financial costs are only the beginning of the problems. Here is how the US Department of Defense describes the global impact of climate change in the present and into the future: "In worst-case scenarios, climate-change-related impacts could stress economic and social conditions that contribute to mass migration events or political crises, civil unrest, shifts in the regional balance of power, or even state failure."[99] The costs of global upheaval due to climate change will reach our shores and in ways that would not be expected simply because Canada experiences a warmer-than-average year.

Those are some of the reasons why the "Canada Is a Cold Country" trope is so insidious. The fact that we are cold does offer us some protection from the worst effects of rising temperatures, but climate change also offers us melting permafrost, rising sea levels, increased heat waves, and more *fire weather*. It likely also means more frequent and stronger coastal storms and perhaps more flooding events. Being a cold country offers us no protection from many of these effects, nor from the spillovers of the costs climate change will bring to the global economy.

I have left out dozens of other effects I could have discussed, including the northward migration of deer ticks that carry Lyme disease,[100] stink bugs and other crop pests,[101] and poisonous snakes.[102] Being a cold country has, in the past, kept some of these threats at bay, and it takes more than a single warmer-than-average year for them to affect us. In a climate-changed world, we will no longer be as protected from the damage that they can create. I've talked a lot about frozen ground in the North, but we face substantial risks in Western Canada from the loss of mountain glacier ice, including far less predictable flows in many of our most important rivers.[103] I have also left out a host of global threats to Canada that are likely to be amplified by climate change. As the Canadian Security Intelligence Service wrote, "climate change presents a complex, long-term threat to Canada's safety, security and prosperity outcomes."[104] And these effects too would not be seen in a single warmer-than-average year.

We should not take solace in studies that suggest Canada is going to be relatively well-off and perhaps even better off with a bit of warming. We can adapt to reduce the impacts of some of the negative consequences of a warming world, but we will not be able to fully eliminate the suffering that a changed climate will bring us. The only way to mitigate the effects of climate change on Canada is to mitigate the effects of climate change for everyone.

4 | Canada accounts for less than 2 percent of the world's emissions

> Last year [the Parliamentary Budget Officer] released a study on "Global Greenhouse Gas Emissions and Canadian GDP" that concluded, "climate change has—and will continue—to negatively impact the Canadian economy." . . . The key phrase is 'global emissions' not Canadian emissions, which total 1.6 percent of global emissions, making it absurd to argue a carbon tax in Canada is going to have any significant impact on global emissions, much less the weather in Canada, up to 77 years from now. That depends on what happens globally.[105]

I do not think I could have found a better quote to lead this chapter than the one above from a June 2023 column by Lorrie Goldstein in the *Toronto Sun*. Goldstein acknowledges that climate change has and will continue to cause significant hardship in Canada but pivots to how we are foolish to consider doing our part to reduce global emissions because we account for less than 2 percent of the world's emissions. Goldstein follows that with a reminder to his readers that reducing climate change damages in Canada depends on others making different decisions than he would recommend for us. It's perfect.

Goldstein is no outlier; you will often hear the excuse that, while mitigation might be desirable in theory, Canadian efforts to mitigate climate change will be like "emptying a pool with a soup spoon, as someone else fills it to overflowing with a firehose."[106]

Feelings of futility have permeated analysis of Canada's climate change mitigation efforts for decades. A recent report from the same Parliamentary Budget Officer quoted in Goldstein's column relied on analysis that "does not account for the benefits of reducing Canada's greenhouse gas emissions in terms of reducing the economic costs of climate change" because Canada's emissions "are not large enough to materially impact climate change."[107] If every *small* emitter adopts similar analysis of the benefits of climate change mitigation, we will not stand a chance.

There is no real dispute that our emissions are small in the global context. Despite growth in our domestic emissions since 1990, rapidly growing emissions in the developing world have caused Canada's share of global emissions to drop from a little over 2 percent in 1990 to a little over 1.5 percent today.[108] By simple math, Canadian emissions reductions cannot solve climate change. China, which accounts for roughly 25 percent of global GHG emissions, is perhaps large enough to consider how its policies will affect the trajectory of global climate change.[109] After that, emitters get pretty small, pretty fast. The next largest emitter to China, the United States, has about half the emissions. The next largest after that, India, has half the emissions of the United States. Indonesia and Brazil are about half of India's emissions. Divide their emissions in half again, and there is Canada, twelfth on the list of emitters.

Combined, the top 12 emitting countries account for about 80 percent of global emissions. None of these countries, even China, is a large-enough emitter that it can *solve* climate change by acting alone. Climate change mitigation is a collective action problem. But we should not minimize our capacity to influence global action, nor should we dismiss domestic motivation for Canada to reduce emissions. As Andrew Coyne wrote in the *Globe and Mail*, solving the problem depends on *small* emitters like Canada finding reasons to act and bringing others along:

> At 1.6 percent of global emissions, Canada is a bit player, next to the Chinas and the Indias of the world. We could cut our emissions to zero, and it would make next to no difference to the Earth's fate. . . . If every country adopted this line of thinking, and used it as a pretext for inaction, the problem would never be solved. Whereas if every country pitches in, it can be.[110]

Coyne makes a compelling appeal to Canada's collective sense of responsibility, similar to one made by Brendan Frank in a blog for the ecoFiscal Commission:

> If Canada balks at climate leadership claiming that 1.6 percent is not enough to matter, some 180 nations could collectively reason that they are not the problem, either. Imagine our obstinacy, rather than our initiative, becoming an example to the world.[111]

Coyne's column goes on to argue that Canada should seek cost-effective policies to mitigate emissions and that we should avoid specific, binding targets. He sees carbon pricing policies as more attractive than other possible interventions we might consider. I don't disagree, but, as economists, we frequently leap to advocacy for carbon pricing without answering the most important questions: why should we act, and what should inform the stringency of our policies? So, let me start with those questions here.

* * *

Canada's obligation to act starts with our substantial, historical contribution to climate change. Simon Evans estimates that Canada has accounted for 2.6 percent of cumulative global GHG emissions, tenth among the world's countries. Depending on how you measure it, our cumulative emissions rank either first or second in the world on a per capita basis.[112] Canada has caused more than its share of climate change.

Canadians also continue to do more than our share to make climate change worse. We rank fourth in the G-20 in terms of GHG emissions per capita, with only Saudi Arabia, Australia, and the United States recording higher emissions per person.[113] Our per capita emissions are about three times the global average. This is not solely a result of our high energy production and exports. Our per capita emissions are roughly two-and-a-half times the global average on a consumption basis, too, owing to our wealth, our transportation choices, our relatively high building heating and cooling loads, and our reliance on fossil energy sources.[114] No matter how you measure it, Canada is making the climate change problem worse.

While some see our historical and current emissions per capita as reason enough for action, many Canadians need more convincing. That we have high per capita emissions and have emitted a lot historically does not mean that our actions will have an outsized impact on future climate change. For some, calls to collective action will carry the day. Analogies to the war efforts permeate discussion of climate change mitigation in Canada.[115] Like the war, we cannot solve climate change on our own. But, unlike the war, Canadians have yet to feel the compelling case for collective action.

Here, then, is a different call to action:

> Canada (is) obliged to (reduce GHG emissions) because even though we produce only two percent of global GHG, we are a disproportionately large producer of energy and other natural resources. Energy is one of the key drivers of the Canadian economy, and it increasingly defines our place in the world.[116]

That was former Prime Minister Stephen Harper in 2007 during his "clean energy superpower" address to the APEC Business Summit. I do not often find myself agreeing with Stephen Harper on climate change, but there is no question that our role as a major energy producer compels domestic action to mitigate climate change. And, to not do so will make it nearly impossible for Canada to argue credibly that international countries should act to mitigate emissions or argue that their policies should not target our own outsized emissions.

The Canadian economy depends heavily on energy production, and energy markets are global. That leaves our energy industry vulnerable to measures and costs imposed on us by the rest of the world to mitigate climate change. Border carbon adjustments and clean fuel standards have already targeted our emissions-intensive oil sands production. Oil and gas pipelines have faced a decade-long proxy war over climate change.[117] Global oil majors have exited the Canadian oil patch in droves, including near-complete divestiture by Shell, BP, Equinor (Statoil), Chevron, Devon, and Total, and a partial exit by Conoco-Phillips.[118] Similarly, banks, pension funds, insurance companies, and even Norway's sovereign wealth fund have ceased doing business in the Canadian oil industry because of impacts on the environment. The sector's reputation has soured enough that the Trans Mountain pipeline is effectively uninsurable unless the company providing insurance can be promised anonymity.[119] We do not have the option to avoid climate change policy, but we do have the choice of whether to abide by policies of our choosing or by those likely to be imposed upon us by others.

There is no option to return to business-as-usual for the Canadian oil and gas sector. Nor will credible domestic policies shield the sector from proxy fights over climate change, but opposition to any development of our resources will be more substantial if Canada steps back from credible action on climate change. To some degree, we can choose whether the future prosperity of our resource industry (or lack thereof) is guided by domestic or international policies, but there can be no debate as to whether Canada should act to mitigate climate change. We must. The more difficult questions are how aggressive our actions should be, and what form our actions should take.

* * *

I hate the performative exercise of national emissions reduction target setting. The global climate conversation has focused for too long on reduction commitments. At regular intervals, we have seen countries commit to different percentage reductions relative to different baseline years, with their promises to be met, we hope, years or decades into the future. It is an ineffective and economically inefficient process.

The lowest cost climate mitigation strategy is to get as many countries as possible to put in reasonably similar levels of effort. This will be economically efficient if our combined efforts are sufficient to eliminate those emissions which can be avoided by incurring costs less than the social costs of carbon. But effort is hard to measure. Is Canada's original Paris Agreement commitment to reduce emissions by 30 percent from 2005 levels by 2030 more demanding than the United States' commitment to reduce emissions to 26–28 percent below 2005 levels by 2025?[120] Or, how does it compare to the United Kingdom's commitment to reduce emissions to 40 percent below 1990 levels by 2030? While we can easily adjust each of these quantity targets to common baseline years, that will not allow a fair comparison of the effort required to meet each country's targets. And, if some countries are acting much more or less aggressively than others, we are spending more resources than we need to in order to reduce emissions.

Ian Parry and co-authors at the International Monetary Fund (IMF) compared Paris Agreement commitments across countries. Their work shows how inefficient and uninformative the game of competing emissions reduction pledges can be. By evaluating the pledges in terms of the dollar value of the carbon tax required to meet them, they found that the Paris pledges of the G20 countries vary substantially: "nine countries need GHG prices below US$35 per tonne in 2030 to meet [their Paris Agreement] mitigation pledges, another nine countries need prices between US$35 and US$70 per tonne, while 12 countries need prices above US$70 per tonne."[121] Economically, this means that we are leaving inexpensive opportunities to reduce emissions unexploited, while chasing some very expensive ones. And, as we know all too well in Canada, when governments make commitments requiring vastly more stringent policies than those that are likely to be imposed elsewhere, those commitments generally go unmet.

In the IMF study referenced earlier, Canada's oft-critiqued Paris commitment was estimated to have required carbon prices well above US$70 per tonne, while the United States was judged to have made a pledge that could be accomplished with a US$35 per tonne price. The United Kingdom's pledge was only slightly more aggressive than that of the United States and thus also far less demanding than Canada's commitment. Canada's Paris target was pilloried for being too weak. But Canada's pledges were more stringent than almost every G-20 country when appropriately compared on the basis of the policies that would have been required to meet them. That meant a no-win situation for Canada, in which we need more stringent policies than our trading partners, while being seen to be doing less.

Ratcheting quantity-based emissions reduction pledges to match other countries' targets, while paying limited attention to the policies required to meet them has led Canada to overcommit at almost every international climate conference to date.[122] Canada's tradition of over-commitment usually leads to our governments losing ambition once they are faced with implementing the more stringent policies needed to meet our ambitious pledges. We can avoid this if, instead of targets, we commit to the policies we are prepared to implement.

I would much prefer that Canada focus on the stringency of our emissions mitigation policies than promises of outcomes we cannot guarantee. And I would like to see future climate conferences focus on similar policy commitments. In Canada, we should implement policies stringent enough that global emissions reduction goals would be met if our policies were applied globally. And we should push for similar commitments from others: policies, not promises.

Canada's GHG emissions price, at $65 per tonne of CO_2 today and rising to the equivalent of $170 per tonne by 2030, would be sufficient to meet global goals were it applied everywhere. The World Bank's 2023 State of Carbon Pricing report concludes that a global carbon pricing level of between US$61 and US$122 plus inflation by 2030 is consistent with limiting global climate change to less than 2°C. Similarly, Ian Parry and co-authors at the IMF estimated that a global carbon price increasing to US$75 per tonne by 2030 would meet Paris Agreement goals.[123] If the rest of the world adopted Canada's carbon pricing regime, to say nothing of other policies like our Clean Fuel Regulations or our 2030 coal phase-out, the world would be on track to meet global goals to substantially reduce the impacts of climate change.[124]

* * *

I helped guide a commitment to policies rather than targets in Alberta when I served as chair of Alberta's Climate Change Advisory Panel in 2015. Alberta faced a problem much like the one Canada faces at global meetings: we were under pressure to commit to deep cuts in emissions, even where those commitments would require more stringent policies than those in place elsewhere in Canada. In my first press conference as chair, I was asked by veteran Alberta political reporter Graham Thomson what targets I would recommend. My answer was that we were going to worry about setting policies, not targets.[125] That is what we did, and "policies, not targets" is a mantra I have repeated many times since.

Based on our recommendations, Alberta implemented Canada's most stringent carbon pricing regime in 2016 ($30 per tonne, matching the price in

British Columbia), combined with a phase-out of coal-fired power and government procurement of renewable power generation. Alberta also pledged to increase its emissions price further, so long as other provinces did the same. We knew that matching commitments made in Ontario and Quebec to reduce emissions substantially below historical levels would require much more stringent policies in Alberta, and that made no sense economically and would have been politically untenable.

Alberta's 2015 climate change plan shifted the Canadian conversation on climate mitigation in part because of its focus on policies instead of targets and prices instead of quantities. We implemented policies that, if they were to be mirrored nationally and strengthened over time, would allow Canada to meet its targets. The opportunity now exists for Canada to ensure that our policies, if imposed globally, are sufficient to meet global goals. We should impose the policies we would like to see elsewhere and be prepared, as Alberta was, to increase the stringency of our actions as they are matched and exceeded by others. It is a much better approach than getting caught up in *target bingo*.[126]

* * *

With all that said, what policies should Canada implement?

Carbon pricing is the best policy approach to reduce emissions. Carbon pricing is a catch-all for government policies that create a financial consequence for the emission of GHGs. The simplest of these policies is a carbon tax, where an emissions fee is collected on fuel sales based on the emissions that will be created when those fuels are burned. A carbon tax has been in place in British Columbia since 2008.[127] A carbon tax sets the price of emissions, and market actors determine the eventual quantity of emissions. If market conditions are such that emissions are very valuable, a carbon tax will not reduce emissions by as much as would be the case in a weaker economy, or in a case where cost-effective substitutes for carbon emissions are available. Emissions might even grow, although more slowly than would otherwise be the case, after the imposition of a carbon tax. A financial trade-off can also be established by using tradeable emissions permits, or *cap-and-trade* regimes, where the quantity of emissions is restricted by the number of permits available.[128] In a *cap-and-trade* regime, regulation sets the quantity of emissions, and the market determines how expensive emissions need to be to meet that restriction. If few permits are issued relative to market demands for emissions, the market value of a permit will increase, and this increases the financial trade-off between emissions and the value that could be derived from selling (or not buying, for those who do not already possess them) emissions permits. From the perspective of an individual emitter, both systems imply that emissions have a price, and neither is necessarily more effective than the other at reducing emissions.

When emissions pricing is implemented in practice, it is almost never the prototypical carbon tax or *cap-and-trade* programs studied in economics textbooks. Most often, policies are hybrids, combining design elements from policies implemented in other jurisdictions. In Alberta, and subsequently for the federal carbon pricing backstop, I have helped develop a particular type of hybrid carbon pricing policy: a fuel charge for small emitters, which functions much like a carbon tax, accompanied by lump-sum rebates to reverse the regressive nature of a pure carbon tax, combined with a parallel large emitters carbon charge that functions much like a *cap-and-trade* regime, with output-based allocations of emissions permits provided to facilities in order to reduce potential competitiveness impacts on industrial emitters. Combined, these two systems place the same value on emissions reductions across the economy, with the exception of emissions reductions achieved through reduced industrial output.[129] The consumer lump-sum rebates do not alter the value of reducing emissions. For example, you save the full carbon charge as well as the fuel costs if you decide to cycle to work, and this is unaffected by receiving the rebate, but the rebates mean that most Canadians are made better off when considering both their direct and indirect carbon charge expenses and the value of the rebates they receive.[130] I am very proud of helping develop and implement these systems in Canada and continue to advocate for them provincially and nationally.

Economists favour carbon pricing policies to reduce emissions because they require the least information on the part of governments, provide the strongest rewards for innovation, and harness the market to ensure that emissions reductions are achieved in the most cost-effective way. Unlike regulations and subsidies that are costly to implement, carbon pricing regimes can also generate revenue, providing fiscal flexibility to governments. Most importantly, the impact of carbon pricing depends on the price: a $30 per tonne carbon price is never going to convince a person or a firm to take $70 worth of effort to reduce emissions by 1 tonne. The more emissions reductions you want to achieve, the higher the carbon price is going to have to be.

Economists have been criticized for being near-religious about carbon pricing, in particular in the United States, where price-based policies have proven politically challenging to enact or have remained too weak to have real impacts.[131] While carbon pricing is the most cost-effective way to reduce emissions, we should prioritize policies that will actually be implemented. If, as has been argued by many, political constraints mean that high carbon prices won't fly in a particular jurisdiction, but stringent regulations are an option for policymakers, it's a no-brainer from an environmental and economic perspective. Where we have the choice, price carbon at levels sufficient to meet global emissions reductions goals. If we do not have the choice, go to the next best thing, so long as that doesn't mean compromising much in terms of the expected rate of emissions reductions.[132]

Fortunately, carbon pricing has survived many elections in Canada, including federal elections in 2015, 2019, and 2021 and myriad provincial elections in British Columbia and Quebec. Pricing industrial emissions has been the norm in Alberta since 2007. It is politically feasible here, just as it has proven to be in Europe and many other jurisdictions.[133] But, carbon pricing remains under political pressure in Canada in part because we have committed to targets that require either a more expensive carbon price or substantial incremental policies over and above the prices in place today. In the most recent projections Canada submitted to the United Nations (see Figure 1 in the Introduction) current policies will see emissions reduced by only 16 percent below 2005 levels by 2030, short of both our original and updated Paris Agreement targets despite aggressive carbon pricing.[134] Carbon pricing is effective, but it isn't magic.

Despite my economist credentials, I have not advocated solely for carbon pricing, nor would I characterize it as the best option in all cases. On the Alberta advisory panel, I supported additional policies, including a phase-out of coal-fired power and a renewable energy procurement plan. Each of these played an important role in the success of the Alberta Climate Leadership Plan, both because they provided substantial emissions reductions and because they corrected other market imperfections. The negotiated coal phase-out allowed for the possibility of a planned—rather than a chaotic—transition away from coal. Alberta's electricity market does not offer any form of market compensation for long-term reliability, something that is clearly valuable. That long-term reliability is not priced constitutes a market failure, one that might have been exacerbated by an exclusive adherence to carbon pricing. And renewable power procurement allowed the province to optimize the use of existing transmission infrastructure while also revealing information on the now-very-low costs of renewable energy generation in Alberta. The policy mix might not have been perfect, but it delivered more emissions reductions than carbon pricing alone, and accomplished multiple, explicit policy goals of the government.

The success of complementary policies in Alberta, the dramatic emissions reductions from the Ontario coal phase-out, the success of cross-border methane and vehicle emissions regulations, and the green boom being triggered today by the US Inflation Reduction Act should remind us that policies other than carbon prices can have a major impact even though they might do so at higher economic costs than is strictly necessary.[135] When the social costs of carbon emissions are in the hundreds of dollars per tonne, many regulatory and subsidy policies will yield socially beneficial emissions reductions even if they are not the most cost-effective options available.

I am less concerned about how we act than that we act. But if we are going to act, I will continue to recommend pricing carbon whenever it is the best and most practical option.

* * *

Canada has a duty to itself and to the world to act on climate change. At a minimum, we should do that which we would have others do so that we might benefit from reduced climate change damages. We have always been leaders in global environmental pursuits, and our population views climate change mitigation as an important priority.[136] Obviously, we can't promise that Canadian policies alone can mitigate increased forest fire risk, heatwaves, or sea level rise. Nor do we want to conflate the impacts of our real efforts to reduce emissions with the impacts of meeting global emissions targets. While we must act, we should not make commitments to polices, explicitly or implicitly, that place materially higher costs on industrial emissions than those imposed by our major trading partners, since we risk displacing rather than reducing emissions. But there is no option for *business-as-before*. Our choice will increasingly become whether to act responsibly on our own initiative or have standards and policies imposed on us by the rest of the world. We should not let our small share of global emissions dissuade us from being a constructive part of a global solution. We want our initiative, not our obstinacy, to be an example to the world.[137]

5 | The world will always use (enough of our) oil and gas

The majority of global GHG emissions come from the energy sector, and a majority of energy sector emissions come from oil and natural gas use.[138] In Canada, the production and combustion of oil and gas is responsible for the vast majority of our national emissions. The oil and gas industry, globally and here in Canada, has begun to acknowledge the importance of acting on climate change, but has often failed to wrestle with the most crucial implication of such action: meeting global climate goals will mean using much less oil and gas.

The oil and gas industry has responded in various ways to the threat to its business model that action on climate change represents. ExxonMobil, for example, has been implicated in a decades-long campaign of misinformation on the science of climate change.[139] Here in Canada, the oil and gas industry, and governments that have supported them, have responded by highlighting the relatively large emissions from US coal-fired power plants compared to Canadian oil and gas production, by reminding Canadians that oil and gas emissions are a small share of Canada's relatively small share of global emissions, or with campaigns touting the positive aspects of Canadian resource extraction inspired by Ezra Levant's Ethical Oil treatise.[140] The news is not all bad though. The oil and gas industry has periodically supported the implementation of carbon pricing in Canada and, more recently, has made substantial commitments to reduce production emissions through carbon capture and storage in the oil sands. More recently though, as pressure to act on climate change has ramped up, the industry has responded with a new talking point: assurances that the world will continue to use oil and gas far into the future, often buttressed by reminders that consumer products from iPhones to toothbrushes are made from oil and gas.[141]

. . . as we transform our energy development, fossil fuels must remain a part of that mix as there is not a current, viable replacement available. At the end of the day, we cannot go to renewables overnight and the world still needs fossil fuels.[142]

This quote, from the head of a Canadian oil and gas industry association, is typical of what we hear regularly from Canadian oil and gas industry leaders. Striking a similar tone, Cenovus CEO Alex Pourbaix, appearing on CBC Radio's *The Current*, offered that "oil demand globally is higher than it was before the pandemic. And most economists expect that we are going to continue to be growing the use of oil and gas at least over the next one to two decades."[143]

Such optimism over future oil and gas demand is not uniquely Canadian. ExxonMobil, for example, forecasts increases in oil use through 2040 followed by a slow decline to 2050.[144] Likewise, Shell's new CEO Wael Sawan said in March 2023 that he was "of a firm view that the world will need oil and gas for a long time to come" and that "cutting oil and gas production is not healthy."[145] BP is a bit of an outlier among the oil majors, as its outlook offers three scenarios, each predicting peak oil consumption within the next decade.[146] The story is much the same for natural gas, with ExxonMobil forecasting a 25 percent increase in natural gas use by 2050 compared to that today, while BP offers mixed conclusions regarding natural gas use across its scenarios.

So, what's the catch? The catch is climate change. Hidden in outlooks that predict continuing robust demand for oil and gas is the assumption of restrained global action on climate change. For example, ExxonMobil states explicitly that its projections imply that emissions in 2050 will be more than twice the level needed to be on track to meet a 2°C climate change mitigation goal.[147] If you remove the assumption of global inaction on climate change, you are left with a very different picture. For example, BP looks specifically at the impact of aggressive global action on climate change and predicts nearly an 80 percent reduction in oil consumption and nearly a 60 percent reduction in natural gas consumption from current levels by 2050 in a world on track to limit warming to 1.5°C. If the world meets the Paris Agreement goal of limiting global warming to less than 2°C, BP reckons that oil and gas consumption will be roughly 60 and 40 percent below current levels respectively by 2050. Combined, BP expects that the world will use half as much energy from oil and gas in 2050 as it does today if it pursues a 2°C climate goal and a quarter as much if it pursues a 1.5°C climate goal.

The oil and gas industry knows that the world will still use some oil and gas, but far less of it if global climate change mitigation strategies are implemented.

So why do they leave out that last part so often? Are they hiding it from themselves, from the rest of us, or both?

It is understandable that most oil executives and many Canadians skirt challenging questions like how the future of the oil and gas industry might be affected by action on climate change. In his book *Thinking, Fast and Slow*, Nobel-Prize-winning economist and psychologist Daniel Kahneman translates for a general audience his academic research into how humans make decisions.[148] One of Kahneman's subjects is the availability heuristic, a description of how humans, when faced with a difficult problem, pivot to answer an easier question. For example, people might substitute the hard question of "how will climate change disrupt our oil and gas industry" with the easier question of "will the world still use oil and gas in the future?" The answer to the second question is almost certainly "yes." But that provides a false sense of security because action on climate change, sooner or later, will fundamentally disrupt the oil and gas industry.

We cannot know exactly what a world acting on climate change will look like, nor can we predict how aggressively the world will act. Energy and climate models and scenario analysis can tell us what may occur with different levels of action on climate change. Scenarios, even some published by the

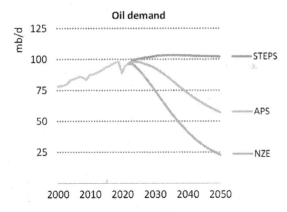

FIGURE 3 | Global oil consumption (demand) in the IEA's World Energy Outlook, 2022.[149] The three scenarios shown are: Stated Policies (STEPS), which models policies currently implemented, equivalent to roughly 2.5°C of warming by 2100; Announced Pledges (APS), which assumes all government greenhouse gas targets are met as pledged, resulting in roughly 2°C of warming by 2100; and a Net-Zero Emissions by 2050 scenario (NZE), which is consistent with limiting global warming to 1.5 °C by 2100. Graphic used under CC 4.0 by licensing.

world's major oil and gas producers, show clearly that increasing oil and gas use is inconsistent with aggressive global action on climate change, and that's despite potential emissions reductions from technologies like carbon capture and sequestration.

The scenarios published each year in the IEA's *World Energy Outlook* show clearly how oil and gas markets could be shaped by the speed and stringency of global action on climate change.[150] For example, in their 2022 *World Energy Outlook*, the IEA describes a plausible transition to net-zero global emissions in which global oil consumption, shown in Figure 3, declines from 94.5 million barrels per day in 2021 to 75.3 million barrels per day by 2030, and drops sharply to 22.8 million barrels per day by 2050. The IEA sends much the same message about natural gas in a net-zero world, predicting a 72 percent decline in fossil gas use by 2050, a drop not quite as sharp as the 76 percent decline in oil use predicted over the same period. The IEA does contemplate more robust oil and gas consumption in their Stated Policies (STEPS) scenario, but this scenario entails global temperature changes closer to 2.5°C by 2100, leading to a markedly different global climate system than in either the Announced Pledges (APS) or Net-Zero Emissions (NZE) by 2050 scenarios.[151] The IEA's conclusions are clear: the world will still use oil and gas if climate change mitigation proceeds successfully, just nowhere near as much of it.

The IEA scenarios are not forecasts. And while the IEA scenarios, like those from BP quoted earlier, offer plausible evolutions of the global energy system in a world acting to different degrees on climate change, they do not represent the only plausible means to achieve any given climate change goal. In the jargon of economists, they provide neither sufficient nor necessary conditions. For example, the IEA net-zero scenario sees oil consumption drop by 76 percent by 2050, but that does not mean that cutting oil use by that amount is required to reach net-zero by 2050, nor does it mean that a drop in oil consumption that large will, in and of itself, assure that we are on a net-zero path. And, similarly, while oil consumption is flat beyond 2030 in STEPS, a scenario that the IEA sees leading to 2.5°C of warming by 2100, this level of warming is not a guaranteed outcome from stable oil consumption over time. The IEA scenarios do provide a sense of the scale of emissions cuts and by extension the scale of market disruption facing the oil and gas sector in a decarbonizing world.

Modeling work done in support of the IPCC AR6 provides a better sense of the range of oil and gas consumption that is consistent with varying degrees of action on climate change.[152] Figure 4 shows a summary of results from energy model evaluations of optimal global oil and gas supply for different climate change mitigation goals under various assumptions about technological progress, population growth, and economic activity. The bad news, for

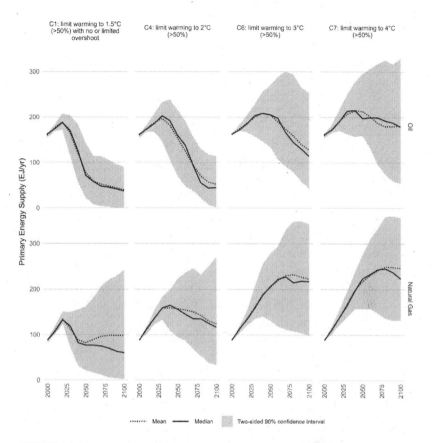

FIGURE 4 | Mean, median and 90% confidence interval for total energy supply from oil (top) and natural gas (bottom) in IPCC AR6 model runs. Scenarios C1, C4, C6, and C7 represent different emissions pathways corresponding to different levels of expected warming by 2100, with temperature exceedance probabilities provided in brackets. Data via Byers et. al. 2022, AR6 Scenarios Database hosted by the International Institute for Applied Systems Analysis (IIASA), author's graphic.[153]

those with a long-term bet on oil, is that there are no modeled outcomes in which aggressive action on climate change is coincident with increased global oil consumption over the long term. With action on climate change sufficient to limit warming to less than 1.5°C, shown in the top-left panel of Figure 4, substantial declines in oil use are immediate and inevitable, with mean and median energy supply from oil dropping to roughly one third of current levels by 2050. If the world follows a less aggressive 2°C mitigation trajectory, shown second from the left in the top row of Figure 4, there is more variation in modeled outcomes for oil consumption, but the model runs eventually point to substantial declines. Only in mitigation scenarios with 3°C or more of warming do we see any real possibility of increasing oil use over long time horizons.

The results from the energy systems models that informed the latest IPCC reports bring some better news for those betting on natural gas. The range of plausible global gas use coincident with aggressive action on climate change is much broader. Even for the most stringent action on climate change considered, shown in the left-most plot in the second row of Figure 4, some models call for more reliance on natural gas in the future to enable a rapid shift away from coal and oil. But, there's a catch. The outcomes that combine aggressive action on climate change and increasing natural gas generally also include substantial deployment of carbon capture and sequestration to reduce emissions from natural gas use. Carbon capture can play much more of a role in mitigating emissions from natural gas use than it can for oil use because natural gas is used more often in large, industrial facilities which are more amenable to carbon capture than a moving tailpipe. The conclusion remains: there is no future for growing, unabated natural gas emissions in a world acting aggressively on climate change.

Action on climate change is also not the only force disrupting global oil markets. Technological progress, in particular the widespread electrification of transportation, means that oil consumption is likely to peak in the near term. In its recent Oil 2023 report, the IEA expects that global consumption of gasoline has already peaked and that oil use for transportation will peak in 2026.[154] Petrochemical demand, the link between oil and iPhones and toothbrushes, is expected to continue to lead to small annual increases in oil consumption through 2028, but given that petrochemicals represent less than 20 percent of current oil consumption, that will likely not be enough to compensate for decreasing demand for oil from transportation by the end of the decade. The world is going to use less oil no matter what actions we take on climate change, and much less if we take aggressive action to mitigate GHG emissions. Similarly, for gas markets, renewable electricity and potential advancements in nuclear power limit growth in demand, even in less emissions-constrained scenarios. A focus on reduction of methane emissions also provides additional fuel for combustion than would otherwise be available, while advances in biogas limit growth in demand for traditional, fossil-derived methane. These changes will amplify the impacts of climate change mitigation on global oil and gas demand.

* * *

Stringent action on climate change will mean less demand for oil and gas, but that does not necessarily imply lower prices nor that the value of Canada's oil and gas production will be compromised. Scenarios like those from the IEA can provide a false sense of confidence that we can predict what will happen to prices in a world acting on climate change. As an economist, it pains me to remind people that oil and gas consumption and prices do not always move in the same direction. In the past, higher consumption has not always meant higher prices; in the future, lower consumption may not lead to lower prices nor to declining asset values. Yes, it is true that if demand decreases in the graph that every student learns in Economics 101, prices drop. But that simple, graphical manipulation assumes that all else is equal. There is another curve on that graph, the supply curve, and what happens to global supply in response to potential action on climate change will be as important as demand changes in determining future prices.

Oil supply at any given time is the product of decades of sustained investment in production and transportation infrastructure. In Canada, we produce roughly 5 million barrels per day of oil, with some from projects more than fifty years old and some from wells less than fifty days old. The global market is much the same. How much investment in production capacity will be made in the coming years depends on what producers expect regarding future oil prices and emissions policies and, in some markets, the willingness of banks, insurance companies, shareholders, and governments to enable further production.

If you do not buy my Econ 101 explanation, we have two recent examples of rapidly falling prices coincident with increasing consumption: natural gas prices from 2009 onward and the more recent drop in oil prices beginning in 2014.

In summer 2008, natural gas at North America's most important trading market, Henry Hub in Louisiana, was selling at over US$13 per million British Thermal Units (mmBtu). Within a year, short-term prices had dropped by more than 50 percent, with long-term futures prices also down substantially. North American prices have never regained their 2008 strength, and, as of this writing, gas is trading around US$3 per mmBtu, despite substantial increases in consumption in North America. Similarly, global oil prices dropped from over US$100 per barrel in 2014 to less than US$30 per barrel in 2016 despite continued growth in consumption.

In each of these cases, production innovations made more resources available at lower prices. This is a rightward shift in the supply curve on that famous Econ 101 graph, and it produced decreasing prices coincident with increasing consumption.

Climate policies could have an analogous effect, leading to higher prices coincident with declining consumption. If producers expect a decrease in future consumption and revenues, they will invest less in developing new reserves, eventually decreasing supply on that Econ 101 graph. Even as demand decreases, the coincident decrease in supply could lead to higher prices. As the saying goes, the surest cure for low prices is low prices (or the expectation of low prices). The more companies reduce investment due to low future price expectations, the higher prices are likely to be.[155]

Just how much are energy producers investing? Global oil and gas investment peaked at close to US$800 billion per year in 2014 and has not been above US$500 billion in any year since.[156] Producers cut their share of capital investment from 90 percent of cash spending in 2016 to less than 50 percent of cash spending by 2022.[157] Instead, as mentioned, they are prioritizing debt repayment, dividends, and share buybacks. The oil and gas industry is not betting on the need for substantially increased production capacity.

Investment in oil and gas production will still be needed, even in the most stringent of climate action scenarios. Without spending on new wells, oil and gas production will fall faster than projected in the IEA's modeled transition to net-zero.[158] Prices could spike dramatically, which could lead to energy security concerns that could, in turn, derail action on climate change, a dynamic we have recently witnessed subsequent to Russia's invasion of Ukraine.

While the oil and gas industry has cut its bets on the need for future production, it is still investing based on far higher expected oil consumption than is likely in a world with robust action on climate change. The IEA finds that current levels of fossil fuel investment are more than double what would be needed in a world on track for net-zero emissions by 2050.[159] The capital investment we are seeing today does portend lower prices if aggressive action on climate change comes to pass. And, if those lower prices do come to pass, some of those capital investments may end up as stranded assets.

We have also seen substantial increases in investment in terminals to support the export of natural gas. This has slightly different long-term implications in that LNG terminals enable trade and are distinct from investment in new production. LNG terminals increase the availability of gas in high-priced regions while broadening markets and presumably increasing prices for producing regions and so should increase both production and consumption, all else being equal. As with investment in oil production, cumulative LNG export capacity is now predicted to be much higher than needed to support the volumes of natural gas trade contemplated in the IEA's scenarios for aggressive action on climate change.[160] Here, too, the oil and gas industry seems to be betting on a future with more natural gas consumption and trade than would be expected with stringent actions to mitigate climate change.

How, then, might oil and gas prices evolve in a world acting on climate change? Consistent with the over-investment detailed earlier, the IEA oil consumption scenarios shown in Figure 3 are accompanied by price projections, shown Figure 5, which hold that inflation-adjusted oil prices will drop dramatically over time in a world acting aggressively on climate change. In their net-zero by 2050 scenario, which implies a 40 percent decrease in global oil consumption, the IEA predicts a 65 percent decrease in oil prices by 2030, with further decreases beyond 2030 to prices equivalent to US$24 per barrel in today's dollars by 2050. Those prices are one-quarter of what the IEA predicts prices would be if the world took a less aggressive stance on climate change. The world is still using oil in the IEA's net-zero scenario, but that does not insulate the oil and gas sector from disruption from global action on climate change.

Natural gas is more complicated because the costs of shipping natural gas around the world are much higher than is the case for oil, so regional price disparities tend to be much larger. Within regions, the IEA estimates substantial impacts of action on climate change on natural gas prices, with prices about 50 percent lower in net-zero energy scenarios than in scenarios where energy policies track closer to the status quo. Importantly for Canada, this would mean

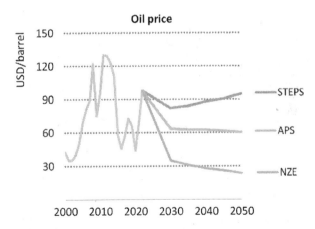

FIGURE 5 | Global benchmark (Brent) oil prices in the IEA's *World Energy Outlook*, 2022.[162] The three scenarios shown are the same as those in Figure 3: Stated Policies (STEPS), which models policies currently implemented, equivalent to roughly 2.5°C of warming by 2100; Announced Pledges (APS), which assumes all government greenhouse gas targets are met as pledged, resulting in roughly 2°C of warming by 2100; and a Net-Zero Emissions by 2050 scenario (NZE), which is consistent with limiting global warming to 1.5 °C by 2100. Graphic used under CC 4.0 by licensing.

lower margins for prospective LNG terminals in a world acting aggressively on climate change. The IEA expects LNG trade to drop by two-thirds by 2050 in a world acting aggressively on climate change, raising the prospect of stranded liquefaction and re-gasification terminals in a climate-constrained world.[161] As with oil, the world will still use natural gas for decades to come, but that does not mean that Canadian exports will be attractive in a world acting on climate change.

* * *

What does all this mean for Canada? It is one thing to note that Canada will almost certainly produce oil and gas for decades to come and quite another to ask how much we will produce and how valuable that production will be.

A 2023 report from the Canadian Energy Regulator (CER) projects that, in a world acting aggressively on climate change, Canadian oil and gas production and revenues would drop precipitously. For oil, as shown in the top panel of Figure 6, the CER found that a continuation of current climate change policies would lead to a peak in production around 2040, at more than 1,000 cubic metres (roughly 6 million barrels) per day. More stringent domestic and global climate change action would dramatically alter those forecasts. With policies to drive Canadian emissions to net-zero, production peaks earlier and falls below today's levels by 2040. But, the real kicker is global action on climate change, which would dry up export markets. The CER expects global action to meet 2050 net-zero goals could lead to peaks in Canadian production before 2030 and a drop to less than one-quarter of current levels by 2050.

The story is, in some ways, even more stark for natural gas, as shown in the bottom panel of Figure 6. Under current measures, the CER sees Canadian gas production increasing steadily for decades, with faster production growth as time goes on. Canadian action on climate change sufficient to drive emissions down to net-zero by 2050 disrupts that and leads to a peak after 2030 and a subsequent sharp decline to levels half as large in 2050 than they would be otherwise. And, in a world acting aggressively on climate change, the CER sees Canadian gas production declining immediately, and dropping to levels around one-third of current production by 2050.

Canadian policy matters to the future of the Canadian oil and gas industry, but global policy matters a lot more.

The CER 2023 Energy Futures scenarios lay bare the challenges facing Canadian petroleum producers in a world acting on climate change. The industry is the largest and fastest-growing source of GHG emissions in Canada, and long-term projections put these emissions on a collision course with our

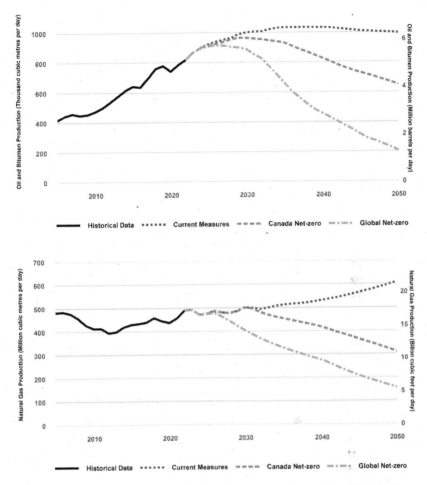

FIGURE 6 | Canadian Energy Regulator 2023 scenarios for Canadian oil (top) and gas (bottom) production under three scenarios: Current Measures, Canada Net-zero, and Global Net-zero. Data: Canadian Energy Regulator; author's graphic.[163]

ambitious national climate commitments.[164] But, much more important is the fact that global oil and gas use must decline in order to meet climate mitigation goals, hence the challenge facing Canada's oil and gas producers and the governments that depend on energy for revenue and economic growth.

It is not surprising that the prospect of a rapid shift away from fossil fuels is unsettling for oil producers and people who live and work in oil-producing regions of Canada. The Western provinces, in particular, have benefited handsomely from oil and gas extraction for decades. In most recent years, more than half of Canada's resource wealth has been tied to oil and gas.[165] People are not merely worried about the potential loss of future oil and gas sales. Large-scale investments in oil and gas production, particularly in the oil sands, result in many thousands of jobs and endless spinoffs into the broader economy.[166] Without those investments, the whole of economic life in Western Canada is different.

Oil and gas prices are reasonably high today, and energy revenues are booming, but despite its bullish rhetoric, the global oil industry is no longer certain enough about its long-term future to bet on itself. Cash flows have been returned to shareholders rather than reinvested in future production. Investment in relatively expensive; long-term projects like the oil sands has declined more than average.[167] Western Canada has been hit hard.

There is likely no going back to the *good old days* even if the world fails to act on climate change. There is no doubt that the world will always use (some) oil and gas, and perhaps even more in the short run than we're currently consuming, but the oil and gas industry, the western provinces, and the country have yet to come to terms with the long-term reality of their situation. When we tell ourselves that the world will continue to use oil and gas, we are expressing hope the Alberta and Canadian oil and gas economy that we have known to date will remain robust in a world acting on climate change. We are likely setting ourselves up for a nasty surprise.

* * *

While climate change risks affect Canada's petroleum sector as a whole, the most substantial economic implications will result from the impacts of climate action on the Canadian oil sands industry, which currently accounts for the majority of Canadian oil production and about three-quarters of the estimated value of Canadian oil and gas reserves.[168] The CER projects that oil sands production will decline rapidly in a world acting aggressively on climate change, dropping roughly 80 percent from current levels by 2050. I would argue that this is far from guaranteed.

The oil sands outlook in the CER's 2023 report provides the latest salvo in a rising debate among Canadian banks, academics, think tanks, and advocates over the future of the oil sands industry in a carbon-constrained world. Just as the question of whether the world will continue to use oil into the future allows us to avoid the more challenging and complex questions of how action on climate change will disrupt the global oil market, debate over the future of the oil sands has come to be framed around whether the last barrel of oil produced might come from Alberta. The answer to that question is almost entirely irrelevant and distracts from important questions of the future viability of the oil sands industry in a world acting on climate change.

> Market forces will not eliminate legacy oil sands production before other sources. There is a reasonable argument that, if there is a last barrel of oil produced in North America, it will come from [the oil sands] unless government policy decides to actively forgo that economic opportunity.[169]

In an excellent contribution to the *last barrel* debate from which the above quote is drawn, the University of Calgary's Kent Fellows argues convincingly that the low variable costs of the oil sands projects already operating in Alberta mean they will be around for longer than most other oil-producing assets.[170] Most people think of oil sands as expensive because they've seen figures for the all-in costs of new projects, which include the recovery of relatively high up-front capital costs, but once those costs are sunk, oil sands production is surprisingly inexpensive. On the contrary, Aaron Cosbey with the International Institute for Sustainable Development argues that the oil sands are vulnerable to global price declines that he is sure will follow as oil consumption peaks and declines.[171] Like Cosbey, the latest report from the CER does not conclude so much that the oil sands are vulnerable to action on climate change but rather that they are vulnerable to sustained low oil prices. The CER relies on oil prices from the IEA scenarios shown in Figure 5, and so they assume that prices will be one-third as high in a net-zero world as in a world taking less aggressive climate action. While Fellows' conclusions might seem at odds with Cosbey's and those of the CER, they are not. Fellows considers the possibility that government policies may, either directly or indirectly, lead to an earlier shutdown of oil sands operations than markets unimpeded by climate action would dictate, which is more or less what both the CER and Cosbey assume will happen in a world acting on climate change. Who produces the *last barrel* is not the most important question. Prices matter more than anything else.

The future of the oil sands depends, for the most part, on whether the world oil market is willing to pay high *enough* prices for oil and, to a lesser degree, on whether domestic climate policies increase the costs of production and transportation. With governments proposing more aggressive carbon pricing, a clean fuel standard, an oil and gas emissions cap, and clean electricity regulations, there is a substantial likelihood that oil sands production costs will grow faster than prices, compromising some of the profitability that is key to Fellows' argument if oil prices are relatively low. If oil prices remain relatively robust, the oil sands industry is in a position to weather significant demands from domestic climate policy and will have the capacity to invest in costly emissions abatement technologies to defray some of the impacts of higher carbon prices.

On the whole, I tend to agree with much of what Fellows finds in his paper: the oil sands are relatively low cost and offer consistent production over a much longer cycle than many other oil investments. Because of these characteristics, Canadian oil sands production did not fall as fast as that in other regions in the wake of the 2008 financial crisis, after the price crash of 2015, or during the COVID-19 slowdown. And, for many of the same reasons, existing oil sands production is likely to be more resilient to changes brought about by domestic and global action on climate change than you might think. Many perceive Canadian oil sands production to be much more precarious than it actually is; a perception that is often encouraged by an industry looking to thwart potentially expensive environmental or fiscal policies. But I also agree in principle with the CER as well as Cosbey: the principal risk to the industry in the long term is low oil prices, a risk that is exacerbated by aggressive global action on climate change, although I must emphasize again that low prices are not guaranteed in a world acting on climate change.

Domestic carbon pricing is an important factor in the viability of oil sands production, but far from the most important. Even though oil sands production generally has higher emissions per barrel than other sources of crude oil,[172] oil sands projects likely fare better under carbon pricing than you think. The relative viability of oil sands production under carbon pricing is best seen by looking at some real-world data. I will use data for the Surmont oil sands production facility, recently purchased from Total by Conoco, to develop an illustrative example. Surmont averages approximately 0.06 tonnes of CO_2 equivalent GHG emissions per barrel produced, on production of 138,000 barrels per day.[173] In 2023, carbon prices in Alberta were $65 per tonne of CO_2 equivalent, which would imply a potential emissions cost of $3.90 per barrel produced. But do not stop there. Under Alberta's and Canada's emissions policies, industrial emitters receive free allocations of emissions credits with each unit of production, which reduces the average cost of the policies.[174] Under Alberta's 2023 emissions policies, oil sands facilities receive sufficient emissions credits to offset 86% of their historic emissions per barrel, which trims average carbon emissions costs to a little over 50 cents per barrel. And do not stop there, either. Emissions costs are deductible from taxes and royalties, bringing the after-tax cost of emissions closer to the "Timbit per barrel" value economist Dave Sawyer once cited.[175]

Canadian GHG policies are slated to become much more stringent, both due to increasing carbon prices and reduced allocations of free emissions credits. But it would take either a substantial further increase in domestic GHG policy stringency beyond those already announced or a big drop in oil prices to render a project like Surmont unprofitable. Based on my own calculations using recent Alberta government data on oil sands revenues and costs, a project like Surmont would continue to operate at a profit even with average carbon emissions costs over $500 per tonne. How can this be? It's simply the upshot of current record high profits per barrel produced and, by extension, per tonne of GHG emissions. In 2021, the most recent data published by the Alberta government, Surmont's average operating and capital costs were $15.42 per barrel, to which the government added $4.61 in royalties.[176] These were set against $53.46 in revenue per barrel of bitumen produced, leading to an operating margin of $33.43 per barrel. Dividing by emissions per barrel, that equates to $557.16 in pre-tax operating profits per tonne of GHG emissions. And revenues have been higher in 2022 and 2023 than they were in 2021, further increasing profits and resilience to GHG emissions policies.

More importantly, a project like Surmont would remain profitable under currently legislated carbon pricing policies so long as benchmark light oil prices (West Texas Intermediate) remain above roughly US$35 per barrel plus inflation.[177] And all of this is assuming that project proponents choose to pay ever higher GHG emissions charges rather than invest in potentially more cost-effective ways to lower emissions, including carbon capture and sequestration. Absent new, costlier emissions policies, oil sands projects will remain valuable as long as oil does. This resilience to future oil and carbon prices may explain why Conoco was willing to pay more than $3 billion for a 50 percent interest in the Surmont project in May 2023. To make the math work on that valuation, they would have to expect higher oil prices in a world acting on climate change than the CER and the IEA do. Or, perhaps they are simply not expecting climate action to be that aggressive, or expecting that any global climate action won't drive oil prices down.[178]

The results above validate the conclusions in the CER *Energy Futures* report. To approximate a net-zero world by 2050, the CER adopts the IEA's long-term oil price assumptions shown in Figure 5, in which prices drop to the equivalent of US$25 per barrel. Under those assumptions, the CER projects rapid declines in oil sands production. These prices are well below most estimates of break-even prices for oil sands projects, and would see most any oil sands project losing money every year after 2030. At prices that low, I would also expect to see oil sands production drop precipitously, as many projects would no longer be worth operating and would shut down early. On the other hand, when the CER considers the continuation of current policies, including planned increases in carbon prices, they find that oil sands production grows between now and 2050, driven by West Texas Intermediate oil prices which are assumed to remain above $70 per barrel plus inflation. These findings align closely with my analysis of oil sands project resilience to already-announced climate policies presented earlier as well. Domestic emissions policies matter, but oil prices matter a lot more.

* * *

Two other factors might change my broad conclusion with respect to the resilience of existing oil sands projects to action on climate change. The first is that more stringent emissions policies, including a proposed oil and gas emissions cap and recently-implemented Clean Fuel Regulations, could add substantial costs on top of the existing carbon pricing regime and could compromise the financial viability of existing oil sands projects, especially in lower oil price scenarios.[179] These policies might convince operators that investments in emissions reduction technology are warranted, but the prospect of low future oil prices or even more stringent future policies could dissuade them from tying up more capital in a long-lived oil production project.

The second, related challenge, is that it has become much more difficult to finance and insure oil sands projects and related infrastructure. Global capital has been voting with its feet and leaving the oil sands. Some major banks and investment houses, including HSBC, Quebec's Caisse de Depot, and Norway's sovereign wealth fund, have ceased doing business in the oil sands because of impacts on the environment. Most major global energy players, including Shell, BP, Equinor (Statoil), Chevron, Devon, Total, and Conoco-Phillips have also reduced or sold their oil sands positions in recent years.[180] Climate change is not the only factor driving these decisions, but it surely played a role. On the day Shell announced its decision to sell its oil sands assets after years of shareholder pressure, it also announced that executive pay would be tied to performance on GHG emissions.[181]

Action on climate change is affecting the ability to insure oil-sands-linked assets as well. Global reinsurance giants including Swiss Re[182] and Munich Re[183] have begun aligning their portfolios to avoid climate risk exposure, and industry leaders have come together with a focus on "using our underwriting, claims, and risk management practices to help ensure and enable the transition to a resilient net-zero global economy."[184] For one example of how this is affecting the oil sands, Trans Mountain Pipelines, now owned by Canada's federal government, explained in recent regulatory filings that it has already encountered reluctance from insurance companies to insure the pipeline. Trans Mountain argued that any disclosure of a commercial relationship with the pipeline could prejudice the pipeline's prospective insurer, making it more expensive or impossible for Trans Mountain to obtain the insurance it needs.[185] It is going to be much harder to finance and insure oil and gas infrastructure in the years to come, in particular in the oil sands.[186]

Reducing the emissions footprint of oil sands production is existential for the industry. Pressure from government policies, from insurers, from financiers, and from shareholders has made it so. Canadian companies are actively seeking to make the case that oil sands investments are compatible with global and domestic action on climate change, and many companies have made long-term commitments to decarbonize production. Anyone who has turned on a television or opened a newspaper in Canada over the past couple of years is aware that all of the major oil sands producers have signed-on to a goal to see the industry reach net-zero emissions by 2050.[187] Canadians might not be aware that meeting the industry's goal will require massive capital and operating expenditures on carbon capture and storage, and perhaps even small modular nuclear reactors, with a total price tag estimated to be at least $75 billion.[188]

Will it work? And will it matter?

Proposals to decarbonize the oil sands and other large industrial projects through massive build-outs of carbon capture and storage or other abatement technology are not new. Canada's first National Communication on Climate Change to the United Nations in 1994, before the Kyoto Protocol, touted an industry-government consortium in Alberta that had set to work on carbon capture and sequestration.[189] Alberta struck a Carbon Capture and Storage Development Council under Premier Ed Stelmach (2006–11).[190] Alberta's 2008 Climate Change Strategy projected roughly 20 Mt per year of emissions reduction from carbon capture by 2020 and a staggering 139 Mt per year by 2050. Industry previously sought to collaborate through initiatives such as the Integrated CO_2 Network (ICO_2N) and the Canadian Oil Sands Innovation Alliance (COSIA). Confidence in carbon capture was high enough more than a decade ago that the government of Stephen Harper pledged in 2008 to require all new oil sands projects coming online after 2012 to meet emissions reduction standards, which would "effectively require putting into place new carbon capture and storage technologies to prevent the release of greenhouse gases into the atmosphere."[191] Yet, to date, only two major carbon capture and sequestration projects have been built in Alberta—Quest at Shell's Scotford Refinery and the Alberta Carbon Trunk Line, which is linked to the government-backed Sturgeon Refinery—and sequestration rates are less than 2.5 Mt per year, a small fraction of total oil sands emissions.[192]

Why have we not seen more carbon capture and storage deployed to date? In part, it is because carbon capture projects are expensive, long-term investments. But, mostly, I believe it is because a massive build-out of carbon capture asks oil sands shareholders—investors who have already made a long-term bet on oil by owning oil sands company shares—to make a long-term bet on aggressive action on climate change at the same time. If you wanted to bet your money on aggressive action on climate change, would you buy shares in an oil sands company?

On CBC's *The Current*, Cenovus CEO Alex Pourbaix admitted that getting shareholders to back major investments in carbon capture and storage is challenging. Pourbaix explained that money spent on carbon capture and storage and other decarbonization activities "does not come with one penny of revenue. These are entirely costs." He reminded listeners that carbon capture and storage was a thirty-year commitment, and that the business case has to be there "even in the periods where oil is priced very, very low."[193] Otherwise, the projects will not get support from shareholders.

Shareholders will only fund a build-out of carbon capture and storage if oil prices are high enough to justify additional investment to sustain production in the oil sands, if carbon capture is the best option to reduce emissions and cheaper than simply paying any carbon charges that might otherwise result from emissions, and if they feel that the imperative to reduce emissions or pay those carbon charges will remain in place long enough for the carbon capture investment to pay off. If not, governments will need to step in and close the funding gap and/or de-risk future revenues from the sequestration of emissions. These efforts will not be described as subsidies, of course, but as "co-financing arrangements" or carbon contracts for differences or some similar term. I have serious doubts that the oil sands industry will deliver on its emissions reduction promises unless governments pay for almost all of the cost of carbon capture and other emissions abatement technology in one way or another. And, industry seems to agree, as they are asking governments to foot three-quarters of the bill.[194]

Speaking on CBC's West of Centre podcast, Cenovus's Pourbaix framed a potential total commitment of $75 billion over thirty years to decarbonize the oil sands as follows:

> It sounds like a lot of money, but you have to remember right now this industry is (. . .) probably somewhere in the range of about 6-plus percent of Canada's GDP (and) the oil sands industry over the next thirty years would add somewhere in the range of about $3 trillion dollars to Canada's GDP. $75 billion over thirty years (is) a very reasonable amount of capital to protect that kind of GDP creation.[195]

Regardless of how it is labeled, a multi-billion-dollar financial commitment to continued oil production will be a challenge for governments committed to phasing out fossil fuel subsidies, and this will be made even more challenging by the recent record profits earned in the oil and gas sector, to say nothing of the massive losses likely to be booked by our federal government to build the Trans Mountain Pipeline.

It is wonderful to imagine an oil sands industry with a much smaller carbon footprint, but it is going to be a whole lot harder to make it a reality despite decades of government and industry promises. And even if we do succeed in decarbonizing oil sands production, most of the GHG emissions from an oil sands barrel occur when the oil is burned, as the industry has been fond of reminding us for decades. Reducing oil sands production emissions through carbon capture and storage or other approaches does nothing to reduce the impact of the eventual combustion emissions on the climate, and global action on those emissions remains the key threat to the long-term survival of industry. The demise of the oil sands is far from assured it a world acting aggressively on climate change, but its continued success is far from guaranteed.

* * *

While this book is about climate change, the fact that the oil sands also carry a substantial unfunded environmental liability should give us pause when considering the industry's future. The current oil sands reclamation liability is almost certainly in the hundreds of billions of dollars. Despite the scale of the liability, the Alberta government holds very limited financial security to ensure that the required clean-up is completed. Instead, the rules effectively allow companies to use the future value of the oil sands sites themselves as collateral against future reclamation liabilities. Yes, you read that correctly. The large reclamation liability is mostly related to mine sites and associated tailings ponds, but given that three of the four major companies still operating in the oil sands own mining assets, it is fair to call this an oil sands issue.[196] And when combined with the potential implications of global and national action on climate change, it's a recipe for disaster.

How did this happen? It has been the policy of Alberta's energy regulators for decades. Using mining assets as an example, Alberta's Mine Financial Security Program (MFSP) collects an initial, small security deposit at startup, and, as more land is disturbed, larger deposits are collected.[197] The largest deposit held at present for any oil sands mine is less than half a billion dollars, far below what reclamation would actually cost in the event of a default.[198] The premise of the program is that larger security deposits are to be collected as the mines get closer to the end of their production lives. There are also provisions for deposits to be collected earlier if the mine value drops to less than three times the value of the expected liabilities, as might happen in a price crash or when a policy change compromises the profitability of mining operations. A similar regime exists for *in situ* operations, although with different triggers for security deposits based on the overall financial health of facility owners and operators.

That sounds fine until you think about it for more than half a second.

It gets worse when you learn how the system actually values the producing assets. I'll use oil sands mines and the MFSP rules as an example again. Each year, Alberta calculates the values of each mine assuming the future looks like the recent past: the value of each mine is determined based on the last three years of profits, although this can be adjusted downward if futures markets indicate declining prices for oil over the next three years.[199] Long-term price expectations are not accounted for in valuing the mine, so there is no possibility to consider a future price downturn in assessing the risk that reclamation liabilities will not be covered. And such a downturn is what many predict could occur as the world acts on climate change.

Now imagine if such a downturn were to occur. About a year into any downturn, the Alberta government *might* start assessing the financial security owed by oil sands projects. I say *might* because the Alberta government has previously chosen to avoid demanding further security from oil sands miners when prices have dropped. As Alberta Environment Minister Jason Nixon explained in 2021, "the math [of the MFSP] does not work if you have extremely low prices."[200] Leaving aside the problem of a program that doesn't work in one crucial price scenario, let's imagine that next time will be different. Following its own rules, the Alberta government would put a massive cash call to the oil sands sector, offering operators a choice: put more money into your money-losing project or walk away. By doing so, the MFSP would amplify any future low-price shocks and make stranded liabilities more likely. Or Alberta might again decide that the math doesn't work and ignore the MFSP rules, hoping that operators remain solvent and meet their clean-up obligations. That's hardly more satisfactory.

In the words of Alberta's Energy Regulator, the MFSP was designed to provide "a responsible balance between protecting the people of Alberta from the costs associated with the liability of oil sands development in the event an approval holder cannot meet their obligations, and maximizing the opportunities for responsible and sustainable resource development."[201] In practice, we are all underwriting the risk if oil prices drop dramatically, as they might if the world acts aggressively on climate change.

I have not written as much as I might have about oil sands liabilities, but we should not forget them when discussing the resiliency of oil sands projects with regard to action on climate change. If oil sands projects are not robust to the changes in markets and policies brought about by serious mitigation efforts, Canadians will be left with a massive, unfunded environmental liability coincident with economic calamity. And this could well happen in a world that is still using oil and gas.

* * *

Canadian industry leaders are fond of reminding Canadians that the world will use significant amounts of oil and gas even if the world acts on climate change. While this is true, it should not distract from the fact that the outlook for Canada's oil and gas industry is likely to look radically different in a world acting on climate change. What will matter much more than how much oil and gas are used is how much the world is willing to pay for fossil fuels. So long as the world continues to be willing to pay enough for oil and gas, Canadian production is potentially more resilient than you might think to domestic and global action on climate change. But new production investments of the type that have bolstered the Canadian economy for the last two decades are much less likely to emerge if the world gets serious about climate change. The combination of global pressure and domestic policies could also render some existing oil projects unprofitable. And, if that happens, Canadians could be left with both economic calamity and a large, unfunded environmental cleanup.

There is also no guarantee that global energy prices will decline in a world acting on climate change, and certainly no promise that prices will become more predictable than they are today. Given current rates of oil and gas investment, the most likely future scenarios are either a world acting on climate change with too much oil and gas supply available and thus lower prices or, in the event that climate action slows, a world with too little oil and gas production capacity facing high prices. That the world will continue to use oil and gas to some extent is largely irrelevant.

6 | The sun does not shine at night

As fossil fuel prices climb, activists believe people will shift painlessly to renewable energy sources. But they have made a major miscalculation: renewables are far from ready to power the world. Solar and wind can only work with massive amounts of backup power, mostly fossil fuels, to keep the world running when the wind dies down, the sky clouds over, or night falls.[202]

Variable renewable energy, be it from the sun, the wind, or other sources, is the source of endless controversy. Solar and wind power are viewed as expensive and unreliable by some, including Bjorn Lomborg, author of the quote above, and as a panacea by others. The truth lies somewhere in the middle. Low-cost solar and wind power is revolutionizing the global electricity sector and will play a significant role in further decarbonizing Canada's already low-emissions electricity system (multiple provinces operate at near-zero emissions today). But Canada will face significant challenges if we rely too greatly on variable renewable energy sources to meet our high winter energy demands. The problem with renewable energy in Canada is not sunsets, it is winter. In this case, the fact that we are a cold country matters a lot.

Despite sun sets and cloudy days, solar power is revolutionizing the global energy system. Once described as "by far the most expensive way of reducing carbon emissions,"[203] solar power costs have dropped by as much as 90 percent since 2009.[204] Back then, the much-maligned Ontario feed-in tariff program offered industrial solar energy projects a subsidized rate of $420 per megawatt hour, more than five times consumer electricity prices at the time.[205] Today, the average costs of electricity from a new solar farm in Ontario is about $80 per megawatt hour, with costs for an Alberta-based project closer to $60 per megawatt hour, each well below consumer electricity prices in both jurisdictions.[206] We are living in a different world.

Solar cost declines and the speed of solar deployment have consistently caught much of the global energy intelligentsia by surprise.[207] Energy forecasts have, almost without exception, consistently underestimated solar generation growth and overestimated future costs.[208] Grid operators and policymakers have found themselves flat-footed and unprepared for the rapid expansion. For example, in 2016, the Alberta Electric System Operator (AESO) forecast that only 1,000 megawatts (MW) of solar generating capacity would be installed in Alberta by 2030, with no further expansion through 2037, the last year of their forecast horizon. Five years later, in 2021, not much had changed, with the AESO still forecasting 1000 MW of solar installed in Alberta by 2030 and less than 1,200 MW by 2041.[209] Less than two years after that, in early 2023, the AESO was considering the reliability implications of almost 2,000 MW of solar being installed by 2030.[210] The pace of solar uptake in Alberta has been so fast that the 2023 analysis was out-of-date by the time it was released, as there was already over 2,500 MW of solar generating capacity installed or under construction. In June 2023, the AESO presented a preliminary version of its semi-annual long-term outlook that forecast shockingly high cumulative installations of 4,000 MW of solar capacity by 2026 and 6,000 MW by 2035. These kinds of rapid updates to forecasts have been happening everywhere. A record 200 gigawatts of solar-generating capacity was installed globally in 2022.[211] The rate of global installations is accelerating, too, with annual installations of up to 400 gigawatts per year expected in the near term.[212]

The declining cost of solar power is the most important energy and climate change story so far in the twenty-first century, so you are going to miss a lot if you dismiss it on the basis of sunsets. The rapid cost decline means that, in most regions of the world, cheap energy is now available without the use of fossil fuels, and, in many cases, it is able to be deployed close to demand. Together with the declining cost of battery backup, solar power offers the real possibility that many regions of the world will leapfrog to a renewable-focused economy, decoupling emissions from economic development in ways we would not have contemplated even a few years ago.

Low-cost solar and storage will impact the global economy far beyond the energy sector. Imagine how much will change when you can put a factory in almost any sunny, tropical location with self-supply of clean energy with battery backup. Now consider that you can do so at a cost comparable to grid-delivered fossil fuel power in much of the industrialized world. This will alter global comparative advantage for manufacturing significantly over the coming decades and will change the course of global economic development.

Nations that adapt to this cheap energy will win out; those that cling to tradition will, eventually, be left behind.

Progress is happening quickly for wind power, too, with global installed capacity tripling since 2013.[213] Onshore wind energy is now a mature technology deployed around the world, and is able to compete with other sources of electricity without subsidies, especially where GHG emissions are priced. In Alberta, for example, government procurement of wind energy has generated hundreds of millions of dollars in profits for the government in just a few years.[214] The shockingly low costs of that procurement program also generated substantial private sector interest in cheap wind energy, which has led to the expansion of capacity not dependent on direct government support. In 2018, Alberta listed just under 1,500 MW of wind capacity before the government procurement began.[215] The government procured just over 1,000 MW of capacity, and there is currently over 5,000 MW of total wind power capacity either producing or under construction in Alberta.[216]

Like solar, wind has dramatically surpassed recent expectations in Alberta, although some of this is just catching up to historical expectations. The AESO 2021 forecast saw installed wind capacity exceeding the 5,000 MW of capacity that is installed or under construction today only after 2041, and even a rapid clean-technology deployment scenario only saw 5,000 MW installed by 2036.[217] The AESO is now expecting 7,000 MW of installed wind capacity by 2030, with more growth to follow.[218] The speed of recent deployment is staggering, but, unlike solar, Alberta had planned for more wind integration in the recent past. For example, in 2016, the AESO Long-Term Outlook considered forecasts of 5,600–8,600 MW of wind generating capacity in the province by 2037.[219] We are now back on track to meet or exceed these forecasts, while also adding a lot more solar.

Offshore wind has seen a rapid decrease in costs as well, although that is less relevant for Canada so far as we have not seen substantial development to date. According to recent Lazard analysis, offshore wind can provide electricity at between US\$72 and US\$140 per megawatt hour on average, which is still more expensive than fossil fuel power in many regions of the world but cheaper than natural gas power in regions with high gas prices such as Europe.[220] Offshore wind is still in the early stages of development, so we should expect costs to decline rapidly over time.

* * *

Renewable energy cost declines have irreversibly changed the global clean energy landscape. But cost declines mean neither that the sun will shine at night nor that it will be windy at the times we most need power. That is the grain of truth in Lomberg's rant. Our electricity grid has to be balanced at every point in time, which means that electricity consumed at each instant has to be simultaneously generated in a power plant or discharged from a battery. Neither wind nor solar energy can be reliably dispatched when needed. Important as they are in the push toward a cleaner electricity grid, we need to make sure that we do not ignore their shortcomings, in particular those that affect Canada differently than more southern, warmer locations.

Canada's federal government has targeted a net-zero electricity sector by 2035 and begun a process to implement regulations to help meet that target.[221] We are starting from a good place, as we already have one of the cleanest electricity grids in the world. Among the G20 countries, only Brazil and France boast cleaner grids. The United States' electricity grid averages almost three times the GHG emissions intensity of Canada's grid.[222] Today, hydroelectric generation, especially in Quebec, Manitoba, British Columbia, and Newfoundland and Labrador, accounts for over 60 percent of our electric generation. Nuclear reactors in Ontario and New Brunswick account for 15 percent of our national electricity generation. Wind and solar generation provide a much smaller share, although, again, their market penetration is climbing rapidly.

We also possess substantial potential to expand our clean energy generation. We rank second in the world in hydroelectric potential.[223] Canada's substantial wind resource potential is distributed across the country, with concentrated, high-quality wind resources in the Alberta foothills, in northern British Columbia, in northern Quebec, Newfoundland and Labrador, in the Gulf of the Saint Lawrence, and around the Great Lakes.[224] Canada's solar resources, too, are substantial, although our northern latitude means that we do not generate as much per unit of land as would be the case further south.[225] And our northern latitude also means that installed solar modules generate less electricity in winter months than in summer months in most locations. We also have potential offshore wind, wave, and tidal energy resources, although future development of these alternatives remains speculative.

Canada's impressively clean average electricity generation hides substantial regional challenges. Neither Canada's renewable energy potential nor, as shown in Figure 7, our existing clean electricity generation assets are evenly distributed across the country. British Columbia, Manitoba, Quebec, Ontario, and Newfoundland and Labrador have very clean electricity systems, while Alberta and Saskatchewan, lacking both substantial reservoir hydro and nuclear power, remain highly dependent on fossil fuel combustion for electricity generation. Saskatchewan, an exception among Canadian provinces, has

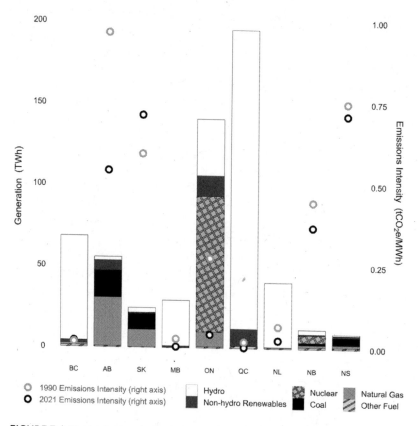

FIGURE 7 | Electricity generation (2021) and 1990 and 2021 GHG emissions intensity of electricity by province.[226] Prince Edward Island, Nunavut, the Northwest Territories, and Yukon each generate less than 1TWh per year and are not shown on the graph. Small quantities of generation from other sources are also omitted for clarity. Fill legend is ordered by column. Data via Environment and Climate Change Canada, author's graphic.

seen its electricity system become more emissions-intensive over the past three decades. Decarbonizing electricity will imply small changes in some parts of the country and substantial changes on the Prairies and in Atlantic Canada. There is an opportunity to integrate cheap, clean renewable power to provide more clean energy in those regions, as well as in regions with existing clean electricity systems. But, to take full advantage of these opportunities, we will need more interconnection of our provincial power grids. Otherwise, we face a more expensive path to decarbonization, and we will have to worry much more about sunsets and calm days.

* * *

Sunsets and calm days matter because energy is not all we need. We also need capacity, or reliable, firm power. We want the lights to turn on when we flip the switch. Solar and wind can fall short in terms of providing reliable capacity, since we cannot dictate when the sun shines and when the wind blows. A few batteries will not be enough. We are going to need to rethink our electricity system at a much more fundamental level to take full advantage of cheap renewables.

Traditionally, electricity systems have been built from the bottom-up. Planners would look at a load profile (how much electricity is likely to be used hour by hour through the year) and plan for two types of generating capacity: *base-load* plants designed to run most hours of the year and load-following or *peaking* plants to provide additional power when needed. When there are substantial economies of scale (i.e., when larger facilities produce lower-cost power), it makes sense to build large *base-load* plants to keep total electricity costs low. Cheap renewables mean that paradigm is shifting. We will still need to supply electricity when it is needed, but a mix of technologies is fast becoming the cheapest way to provide energy and capacity. A report from the Canadian Institute for Climate Choices explains:

> When renewables were expensive, paying a high cost for their intermittent power and having to pay the system cost to integrate them made that pathway a questionable value proposition. Now with recent renewable cost declines, that calculus changes. Renewables plus something—be it peaking capacity, storage, flexible demand, or transmission—to integrate them can now compete with the costs of clean firm power.[227]

A mix of technologies and, likely, more total generating capacity might sound like I am suggesting a more expensive system, but that need not be the case. When talking about electricity, there is a tendency for people to forget operating costs, in particular for coal and gas assets, and to focus only on capital costs or total installed generating capacity. This leads, in turn, to think of building excess generating capacity as a duplication of effort. We should think not of duplication but of substitution, spending on capital instead of on fuel. Consider, for example, that a combined-cycle natural gas power plant's capital costs account for only about one-fifth of the average costs of generating electricity, and its operating costs alone can be greater than the all-in average costs of generation from renewables.[228] And, if carbon capture and storage is added to reduce a gas-fired power plant's GHG emissions, operating costs increase costs even more. This implies that a portfolio of generating resources including, in this example, gas and variable renewable sources can be cheaper over the long term, even though such a system requires building more total generating capacity. Higher up-front capital costs are offset by lower fuel and other operating costs, reducing the total costs of providing low-emissions electricity.

* * *

A portfolio of generating assets can offer lower costs of energy and enhanced reliability at the same time, but the smaller the system, the more redundant capacity you will have to build to make sure that the lights stay on when the sun is not shining and the wind is not blowing. It also helps a lot if the sun tends to shine and the wind often blows at the same times as people want to use electricity.

Building a resilient, renewable portfolio is going to be easier in some parts of the world than others. Solar energy is a great solution for decarbonizing electricity when peak demands occur as a result of hot, sunny days, as is the case in much of the United States. In more northern locations, where energy is needed most when it is cold and dark outside, the challenges with solar are acute. This is especially true on the Prairies, which already face the largest decarbonization challenge in Canada. Wind power provides better geographic diversification than solar—our northern winters tend to be relatively windy— but often the coldest days will bring very little wind.

While declines in the cost of both renewable electricity and battery storage have made it possible to smooth out variability in generation over the course of days or even weeks in a relatively inexpensive manner, seasonal storage remains a substantial challenge.[229] To make full use of renewables, especially solar, you are going to need a lot of seasonal storage in Canada. Even in our best locations for solar, in southern Saskatchewan or southwestern Ontario, about two-thirds of a solar plant's generation will occur between mid-March and mid-September. The panels will generate less than half as much energy in the depths of winter, when we use the most energy, as they do in the height of summer. The issue confronting variable renewables in Canada is not sunsets and calm days, it is winter.

To put a finer point on this, consider Figure 8, which shows the distribution of daily wind and solar generation per unit of installed capacity in Alberta since 2018, for warm versus cold days. The warmest and the coldest days of the year tend to have lower wind generation, which presents challenges for meeting energy demands during those periods even with battery storage.

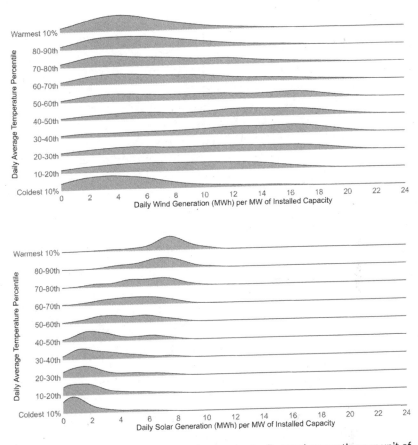

FIGURE 8 | Distribution of daily wind (top) and solar (bottom) generation per unit of installed capacity by daily temperature decile. Daily temperature is measured as the total number of degree hours at the airports in Edmonton (YEG) and Calgary (YYC). Data from the Alberta Electric System Operator and Environment and Climate Change Canada. Author's graphic.

For solar, the relationship is even more stark: the coldest days of the year are much more likely to be to come with very limited solar generation, while the warmest days have plentiful solar energy available. If we want to supply our winter peaks with solar power, we are going to need a lot of long-term storage. The famous Tesla lithium-ion batteries are not suitable for that purpose. And while newer battery designs promise relatively inexpensive longer-term storage, the cost of storing energy over seasons is still likely to be orders of magnitude more expensive than the prices we are accustomed to paying for electricity.

Battery storage becomes more cost-effective as the number of charge and discharge cycles increases, as each cycle allows the costs of the battery to be spread over more units of delivered electricity. This is also why seasonal storage is more challenging: it cycles less often, so the cost of the battery or other means of storage needs to be amortized over fewer units of delivered electricity. In a warm, sunny climate with relatively consistent solar generation throughout the year, batteries can store solar generation from the mid-day solar peaks for use in the evenings. If you can do this most days, perhaps cycling a battery hundreds of times per year over many years, the costs of the battery are spread over the total energy stored and delivered in each of those thousands of cycles, dropping the costs substantially. Whether in a home or industrial application, with frequent cycles, a battery can be a relatively inexpensive tool to integrate renewable energy. Depending on financing and installation costs, costs per unit of electricity stored and delivered are likely in the range of 15 to 25 cents per kilowatt hour, which is within the daily or even hourly fluctuations of consumer or wholesale electricity prices in many markets.[230]

The financials for seasonal storage are dramatically different. Seasonal storage involves shifting power from periods of peak solar generation in the summer to meet peak energy demands in the winter. This application means cycling your battery no more than a few times per year, increasing the costs per unit of stored energy dramatically. Where batteries are used only to provide backup during long periods of low solar and wind generation, they will end up being much costlier per unit of stored energy, all else equal.

Batteries do address some concerns about wind and solar reliability, and they are very useful when generation correlates very well with peak loads, or where smoothing short-term variability is the sole concern, but batteries are not practical for long-duration storage. Unless we are prepared to endure much higher electricity costs and/or much lower reliability, we cannot expect to replace our existing generation fleet with wind and solar generation and battery storage.

Luckily, that is not what is on the table for Alberta or for Canada. We have the luxury of an existing clean electricity grid in much of the country. Our existing and potential new hydroelectric dams can provide both reliable energy and cost-effective storage, while nuclear in Ontario and Atlantic Canada provides substantial reliable supply. To build a cost-effective, decarbonized electricity system across the whole country, we will need to spread out the benefits of those resources nationally, mostly through adding more transmission alongside more renewable and possibly also more nuclear power.

Transmission lines—lots of new wires connected by pylons—are the key to powering an economy using plenty of intermittent wind and solar resources.

Transmission may be the most important and least discussed component of a low-carbon electricity system. People are excited to talk about new solar farms and new wind turbines (well, some people are), but connecting those projects to consumers and allowing those parts of the country without large-reservoir hydroelectric dams or substantial nuclear generating capacity to make use of fuel-saving renewables is going to take new transmission infrastructure. Transmission lines will link those places where the sun is shining and the wind is blowing at any point in time to locations that need electricity, making effective use of batteries and other storage assets and providing each region with backup power. As the cover of *The Economist* read not long ago, if you want clean energy, hug a transmission pylon, not a tree.[231]

Canada has very little intertie capacity (the ability to move electricity between different electricity systems), and remarkably little of what we do have serves the Prairie provinces where our most carbon-intensive electricity grids are located. Alberta's peak electricity demand is almost 12,000 MW, and its combined interconnections could, if all operated at capacity, serve less than 15 percent of that peak.[232] Saskatchewan has a much smaller system with peak demand of roughly 4,000 MW, but its transmission interconnections could meet less than 10 percent of that peak if operated at full capacity. The two provinces can import less than 12 percent of their combined peak load. This will present a substantial challenge to decarbonizing two of Canada's emissions-intensive electricity systems.

Decarbonizing electricity supply is going to require a staggering amount of new transmission infrastructure. For Canada, economists Brent Dolter and Nic Rivers calculated that thousands of megawatts of new transmission capacity are required on the Prairies, along with connections between hydroelectric resources in northern Quebec, and major population centers in Ontario and Quebec as well as to the Atlantic Provinces.[233] The Canadian Energy Regulator reports that interconnection capacity between British Columbia, Alberta, Saskatchewan, and Manitoba should almost triple if we want to deliver a low-cost, net-zero electricity system.[234] In Princeton's Net-Zero America Project, researchers found that a high renewables scenario requires five times as much transmission capacity in the United States as is installed today.[235] Recent analysis from the National Renewable Energy Laboratory estimates that existing transmission capacity must double by 2035 to meet the Biden administration's 2035 clean electricity target.[236] And, many US calculations are made assuming that they have increased access to Canadian hydro to allow them to integrate more variable renewables. We are also likely to benefit from more robust north–south connections in order to access energy from those locations where the sun shines more reliably and for more of the day when it is cold and dark in Canada. Any barriers to or delay in building more transmission will result in a more costly path to decarbonization because we will have to worry more about *dunkelflaute*, the German word for dark, non-windy days.[237]

* * *

It is likely worth a quick digression on nuclear power, especially given its recent resurgence in both public debate and utility system planning in Canada. My thoughts are conflicted. Given safety and waste concerns with nuclear power, I would prefer that we had other options for firm, low-carbon energy, but the value of existing and potentially new nuclear reactors in providing firm power, especially in our cold, dark, energy-demanding northern winters, is compelling. There is a good chance that we will see some new development of nuclear power in Canada if we pursue a path of rapid decarbonization. Ontario is currently exploring a new large-scale nuclear reactor at its Bruce generating station, and as many as four small modular reactors at its Darlington nuclear plant.[238] The Canadian Energy Regulator (CER) projects that nuclear generation in Canada could increase to two-and-a-half times current levels by 2050 under aggressive climate policies, with all new capacity coming from as-yet-unproven small modular reactors.[239] The IEA also sees substantial increases, perhaps a doubling or a tripling of global nuclear capacity as the world acts on climate change.[240]

Nuclear will not be the main driver of decarbonized electricity here or elsewhere; hydro, solar, and wind will fulfill that role. In the CER report cited above, more than twice as much wind and solar generation as new nuclear generation is added under aggressive action on climate change.[241] And, globally, renewable energy will widely outstrip nuclear capacity additions. However, if we fail to make the changes needed to integrate more wind and solar, including new transmission, we will have to choose between more nuclear generation and higher GHG emissions.

Three challenges confront nuclear power: cost, speed of deployment, and flexibility. Nuclear costs have skyrocketed in recent years. In its 2023 study of the levelized (average) costs generation, Lazard estimates that nuclear electricity costs between US$141 and US$221 per megawatt hour, two to six times Lazard's estimates of the costs of solar (US$23 per megawatt hour) and three to six times more their estimated cost of wind (US$31 per megawatt hour) energy. There is substantial optimism over small, modular reactors, but it is hard to see how costs will be substantially lower with a series of smaller units compared to a single, large facility. Comparing energy costs between nuclear and wind and solar is not an apples-to-apples comparison of value, since nuclear plants offer firm power, which is very important for carbon-constrained grids. But nuclear is also more expensive than other firm generation options, including low-emissions options like natural gas plants equipped with carbon capture and storage technology.[242] In its 2023 Annual Energy Outlook, the US EIA sees nuclear energy being outcompeted by cheap renewable power, even in low-emissions scenarios.[243] And Wood-Mackenzie concludes that despite nuclear power's advantages, "the cost gap is just too great for nuclear to grow rapidly," even in a carbon-constrained world.[244]

Another issue with nuclear power is that it does not match well with variable renewables. While there is some global experience in operating nuclear plants flexibly,[245] nuclear and other plants designed for round-the-clock energy provision are more likely to crowd out than complement renewables like wind and solar and vice versa.[246] Nuclear also cannot compete with wind and solar on speed of deployment. By 2025, the US Department of Energy forecasts 400 GW of annual installations of solar per year, and they are probably too cautious.[247] That would add, in a single year, new generating capacity equivalent to the total global installed capacity of nuclear power.[248]

Nuclear is an option; it just is not likely to be our best, fastest, or cheapest option. A material expansion of nuclear generation in North America is unlikely. Rivers and Dolter find that "only when new transmission is not allowed and complete decarbonization of the electricity system is modelled, are new nuclear facilities part of the optimal mix."[249]

* * *

Building a cost-effective grid with more integration of wind and solar and lots of transmission and storage capacity is the best way forward, and while we're building this out, we will also want to change when we use and how we pay for electricity.

For decades, we have spent resources on load-shifting, trying to move electricity demand away from peak periods in order to drive down costs by using larger, *base-load* plants instead of more expensive *peaking* plants for the bulk of our electricity supply. The way many of us pay for electricity reflects this legacy design of electricity systems: time-of-use pricing and demand charges, for example, signal consumers to smooth out their consumption and to use less electricity during peak demand hours when a legacy grid would have to dispatch those expensive, specialized plants. The same rationale, applied differently, will allow us to optimize an electricity system that is more reliant on wind and solar power.

In renewable-dominated grids, peak times or, in more useful terminology, high net-load periods will occur when there is less available supply from variable resources like wind and solar. Adapting to a renewable-dominated grid will mean changing our consumption patterns to shift demand away from potential high net-load periods to periods when wind and solar resources are plentiful. It will mean using energy storage, home-based technology, and even vehicle-to-grid applications to shift demand to different hours of the day. As economist Blake Shaffer writes, if the traditional electricity system relied on dispatching electricity supply to meet forecasts of demand, the modern electricity grid will rely more on "forecasting supply and dispatching demand," with automated home and industrial technology shifting our net electricity demand to periods of plentiful renewable energy. Luckily, some markets already allow for negative prices—yes, you can get paid for consuming electricity—during periods of high renewable energy supply, encouraging these needed changes in consumption. But, this, too, will be more challenging to do in Canada, where these changes involve shifting demand across seasons, not just to different times of the day or the week.

We likely underestimate our capacity to change our behavior to match the type of electricity system we will have in the future. We are used to the system of today. University of Chicago professor David Keith once said something that changed my thinking about the coming energy transition: "it is much cheaper to store water than electricity."[250] Keith was talking about how we should think about energy-intensive processes like desalination of water in a different way, perhaps using cheap, renewable power to desalinate water and, rather than storing electricity to run a desalination process 24/7/365, operating the plant only in hours in which renewable energy is available and storing water when electricity is scarce.

Whether this makes financial sense for desalination is less important than the thought process involved. If we try to build a renewable energy system to replace the system we have today without considering low-cost ways to shift our electricity demands to when energy will be cheapest, we will miss out on a lot of opportunities. The same is true if we dismiss the value of cheap wind and solar power because of concerns about sunsets and calm days.

* * *

This chapter has three conclusions. First, there is a fundamental shift underway in global energy markets driven by the increasing availability of cheap solar power and the declining costs of short-term battery storage. Cheap solar and storage will disrupt the global industrial economy by shifting the locations with the cheapest, reliable electricity toward sunny, equatorial regions.

Second, Canada is bombarded with energy research from the United States, and while much of it applies here, there are times when it does not. The solar power revolution is one of those times. Solar has enormous global potential, but our northern winters' combination of low sun angle, short days, snow and ice cover, and substantial energy demand makes Canada a relatively poor fit for solar power. We will not be able to rely on cheap solar and inexpensive batteries in the same way as much of the United States. That does not mean there is no place for solar; there is absolutely a role for the cheap energy that solar can provide during hot summer days. But we must make sure that we are preparing for a northern winter, not a California summer when considering our future renewable energy mix.

Finally, we cannot expect to simply replace our existing fleet of fossil fuel generation with cheap renewable energy and batteries. This is true anywhere, but especially so in Canada. More reliance on renewables will mean that we have to worry more about planning for sunsets and calm days and that we will need to invest substantially in transmission to minimize our vulnerability to fluctuations in renewable power supply. Furthermore, we will need to change the way we use and pay for electricity to adapt to a more renewable-heavy grid. We are fortunate in Canada to have substantial, existing clean hydro and nuclear generation and the potential to add a lot more of it to complement wind and solar power. We do face a substantial, seasonal challenge that will require either more hydro, perhaps more nuclear, more expensive seasonal storage, or a lot more reliance on electricity trade. Cheap wind and solar power have dramatically changed the global energy system, but we should not underestimate the challenge we face in decarbonizing our power supply.

7 | Governments can plan a just transition

As chair of Alberta's Climate Leadership Panel in 2015, I was asked to provide advice to the government on a number of climate change policy options, including a phase-out of coal-fired electricity. Part of our work included a series of open houses at which Albertans could help set the direction of our province's new climate change policy. Of the hundreds of conversations I had at those sessions, one stands out. There was a long line to speak with me (this does not happen often), and one man stood in line for perhaps forty-five minutes until he finally made it to the front. He walked up, shook my hand, and introduced himself. "I work at the Sheerness coal mine," he said, "and I just wanted you to know that."

He turned around and walked away.

My memory of the event is not perfect, but I swear that I would know that man in an instant if I were to pass him on the street. I thought about that miner's words often during our work and how he, better than nearly anyone or anything else, captured one of the very real consequences of major energy policy shifts. Hidden within the energy models and spreadsheets that economists and engineers use to set policy are real people facing dire consequences.

In today's policy parlance, the plight of that mine worker is captured by a single phrase: the just transition. The phrase communicates a desire to design climate change policies with people like him in mind. But, for me, it raises different questions. It leads me to wonder if we are making false promises to people like that mine worker. I worry that the promise of a just transition gives policymakers and activists permission to ignore some of the real costs of action on climate change. Could I really look that man in the eyes and tell him I was sure he would not be left behind as Alberta transitioned from coal to natural gas, and later to renewable power? Would government programs make him and his colleagues whole? No, I could not, and they would not. More likely than not, we would not be able to compensate for the impact an accelerated shutdown of the Sheerness coal mine would have on his career and his community. There would be a transition, but there was no guarantee it would be just.

I never saw that miner again, but I know what happened to the Sheerness mine and to the two coal-fired generating units nearby. Federal regulations finalized in 2012 would have forced the power plants, originally commissioned in 1986 and 1990, to shut down or convert to natural gas by 2040 so, in 2015, it would have been reasonable for that miner to expect to work for the rest of his career at Sheerness. The 2015 Notley plan accelerated the likely end of mining at Sheerness by committing to a coal phase-out by the end of 2030.[251] Fifteen years is short, but still sufficient time to plan. Then, in early 2017, Sheerness owner ATCO dropped a bombshell: the plants would be converted to gas by 2020, much earlier than previously thought.[252] By March 2021, both Sheerness units had been converted, and the mine is under reclamation. The trajectory above, from the Harper-era coal regulations in 2012, which promised a decades-long planning horizon, through the shutdown of the Sheerness mine took less than ten years.

In Earnest Hemingway's 1926 novel *The Sun Also Rises*, there is an oft-quoted passage about how one goes bankrupt: "gradually, then suddenly." That is what has happened with coal-fired power in Alberta. With a shutdown that was much more rapid than expected, workers like the miner I met at that open house and his neighbours in coal towns were not going to be okay. Government actions provided some cushion for the fall, but they would not and could not fully mitigate the impacts of the sudden exit of a keystone industry for the miners, other residents, and business-owners in Alberta's coal towns.

Contrast the story of the miner I met at our open house with the first key message on so-called *just transition* policies contained in a briefing provided to Minister of Natural Resources, Jonathan Wilkinson:

> Our work on just transition is underpinned by a firm belief that we cannot achieve climate action and the transition to a low-carbon economy without putting people—Canadian workers and industries—first, ensuring no individual or region can be left behind as we move to net-zero.[253]

The briefing explained how the transition to a low-carbon economy in Canada "will have an uneven impact across sectors, occupations, and regions, and create significant labour market disruptions" and warned that "larger-scale transformations will take place in agriculture, energy, manufacturing, buildings, and transportation sectors."[254] This was controversial, but it should not have been. It should have been obvious, as the sectors identified as facing larger transformations account for most of Canada's emissions. What should have been controversial was the government's promise to insulate everyone from harm.

The promise that nobody will be left behind as we move to net-zero emissions is a great talking point, but can governments promise, or even hope to deliver such an outcome? Economic transition will be painful: there will be upheaval, there will be regional pain, and there will be people who never recover. Governments and progressive organizations are quick to promise that a just transition off fossil fuels in the Canadian economy is readily achievable. All we need, we are told, is time and a plan. I do not think that is true, especially in the context of an economy-wide transition to cleaner energy. The promise of a just transition with nobody left behind hides the very real costs of climate policy from Canadians.

Promises of a just transition are going to run into two hard realities. First, the fossil fuel energy industry, and in particular the oil and gas sector, is far larger and more diverse than other industries to which Canadian governments have attempted to provide transitional support. Second, a policy-forced transition away from oil and gas is different from other economic transitions we have weathered in Canada because of the high wages earned by oil and gas workers, the significant government incomes that oil and gas extraction provides, and the very real potential for market signals to counteract government transition planning.

In what follows, I will concentrate on the oil and gas sector since it is the largest challenge in terms of planning any sort of just transition for fossil fuel workers in Canada, but I will draw on examples from a wide range of sectors including coal to illustrate the challenges governments face if they wish to promise a just transition.

* * *

For whom are advocates proposing to plan a just transition? There is no clear definition of a fossil fuel or oil and gas worker, and there is a wide range of estimates of the size of the energy sector in Canada and its impact on the economy.

Natural Resources Canada (NRCan) defines oil and gas workers as individuals involved in the exploration, development, production, and transportation of oil and gas resources. NRCan estimates that there were 163,700 oil and gas workers in 2021, with these workers forming the majority of an estimated of 264,000 total energy sector jobs.[255] Using a broader definition, NRCan estimates that as many as 450,000 Canadians work somewhere along the oil and gas supply chain.[256] TD Economics estimates that between 312,000 and 450,000 workers could be displaced as Canada shifts to a low-carbon economy.[257] These workers account for a relatively small portion of the total Canadian labour force, as some have claimed while promoting plans to phase out fossil fuels.[258] But contemplating a policy-forced transition to different types or locations of employment of anywhere near that many people—the entire population of Victoria or Saskatoon or Halifax—is unprecedented in Canada.

Oil and gas is already an industry in transition. NRCan's higher estimates of 450,000 people employed directly or indirectly in the sector is a decrease from previous estimates of roughly 600,000 for 2019 and more than 740,000 in 2014. Most of this decline is due to less indirect employment: jobs in construction, project management, and finance have been disproportionately affected by the downturn in new project investment in the oil sands and across the sector. Compared to 2014, estimated direct employment is down 15 percent, while estimated indirect employment is down 25 percent. Employment has likely increased since the publication of the latest NRCan estimates, as the oil and gas industry has rebounded post-COVID, although a subsequent series of job cuts in spring 2023 affected thousands more workers in the oil industry in Alberta.

The clean energy transition will create new job opportunities, and some estimates suggest that there will be more total energy sector employment as we transition to cleaner sources.[259] But there is no guarantee that these jobs will accrue to those who bear the costs of displacement. Smaller communities that are heavily exposed to carbon-intensive industries will endure substantial economic costs from the transition, and there is no reason to expect that they will attract offsetting clean energy investment. An energy transition employment shock risks creating some of the same polarizing conditions seen in the United States after the collapse of its manufacturing sector.[260] Canada's manufacturing sector shrunk substantially, too, but this was buffered (or in some minds, caused) by the oil and gas booms of the 2000s. Despite the booming national economy, many in Southern Ontario were left behind in the manufacturing downturn, and many will likely be left behind in any coming shift away from fossil fuel production, processing, and consumption. There is no guarantee that they will all benefit from so-called green jobs.

The macroeconomic and fiscal impacts of oil and gas production are larger than its labour market impacts but are mentioned infrequently in the context of just transition plans. In 2021, while oil and gas accounted for about 1 percent of national employment, it accounted for around 20 percent of the value of Canadian goods exports and roughly 6 percent of GDP, depending on which estimates you prefer. Royalties, lease payments, taxes, and related revenues from oil and gas comprised substantial shares of provincial and national government revenues. It is not easy to replace those kinds of numbers. Rachel Samson of the Institute for Research on Public Policy writes: "Canada will need to find new sources of economic growth, exports, jobs and government revenue to fill a growing hole" as traditional fossil fuel sectors decline.[261]

Canada has endured significant labour market transitions in the past in farming, manufacturing, fisheries, the media, and many other industries. This leads some, notably economist Jim Stanford, to argue that the coming energy transition is not unique, and appropriate government policies and planning can mitigate potential negative consequences.[262] Rather than the fossil fuel industry presenting a unique challenge, Stanford sees the energy transition as receiving more attention because of the alignment of interests of the affected workers with wealthier owners of the companies for whom they work.[263]

I disagree.

The looming Canadian energy transition and the way it is discussed by progressive activists and politicians is uncharted territory. The energy transition will be more challenging than previous transitions in part because fossil fuel workers earn far more than most, and these high-paying fossil fuel jobs comprise more than one-third of total employment in certain regions. It will be more challenging because the fossil fuel sector, while capital-intensive rather than labour-intensive, is a huge wealth and government revenue generator. If, as is highly plausible, domestic government policy forces both employment and revenue to drop in these industries before global markets would dictate such a change, there will be regional tension as we have seldom, if ever, seen before in this country. We have never experienced a policy-forced transition of this sort in Canada, and I do not think we have ever seen politicians and pundits talk so readily about phasing-out high-paying jobs.

In the sections to follow, I discuss why many of the examples that proponents use to assuage fears of the impact of the looming energy transition on Canadian workers are poor fits. We are not preparing for a transition like the Ontario or Alberta coal phase-outs, nor will there be many parallels with the Newfoundland cod moratorium. Similarly, a rapid Canadian transition away from fossil fuels is unlikely to look much like the German coal transition nor the Danish transition away from oil and gas. Each of those transitions occurred in industries that were in decline or in which governments had been active in prolonging the industry well beyond its best-before date. And none of these transitions, absent perhaps the early days of the German coal transition, occurred in industries of the size, scope, and economic importance of Canada's fossil fuel industry today.

* * *

On July 2, 1992, the Canadian government announced a two-year moratorium on northern cod fishing in the North Atlantic. This policy, which followed years of reduced catch limits, placed 19,000 workers in immediate need of financial assistance. The Newfoundland cod fishery transition was not caused by the moratorium; the moratorium was the result of flawed policy that allowed decades of over-fishing. By 1992, the Newfoundland and Labrador population of cod was less than 10 percent of the size it had been three decades earlier, and it will not likely recover in our lifetimes.[264]

There was no business-as-usual alternative to the moratorium in which a viable cod fishery could continue to exist. While there may be no business-as-usual in the oil and gas sector, any near-term Canadian energy transition will not be driven by the physical exhaustion of oil and gas.

There are lessons we should learn from the cod moratorium. First, governments are very bad at planning for a future downturn, even when the signs are clear. Despite the obvious decline of cod stocks, the government denied any need to formulate a transition plan for the inshore fishery until the moratorium was, literally, weeks away. Second, government does not react quickly nor precisely. The initial government response to the cod moratorium suffered from a slow rollout and an inability to identify the appropriate beneficiaries. Finally, it is challenging for government programs to transition workers out of industries or regions. Government support programs brought some fishers back to the industry after they had already left for other jobs and did not dramatically reduce regional dependence on the fishery.[265] Few new industries arose in fishing communities, and most fishers kept fishing.

These lessons are useful, to a point, but the oil and gas industry of today is not the cod fishery of the 1990s. It is easy to imagine governments unwilling to concede that the oil and gas industry is in trouble and lacking a plan to manage a crisis, but that is about as far as any comparison should go. Some draw parallels between calls to reduce oil production to combat climate change and the physical exhaustion of the cod, but we are not running out of oil and gas, and the industry is not in financial peril as the cod fishery was in the 1990s.[266] On the contrary, Canada's oil and gas industry is in the midst of its most profitable years ever, with 2021 revenues of $175 billion dollars and even higher revenues likely to be reported for 2022 and 2023.[267] Oil and gas production both continue to set new records. Government revenues from the oil and gas sector are at an all-time high.[268] Before the moratorium, the Newfoundland and Labrador cod fishery was relatively small; in today's dollars, the inshore cod fishery of 1990 would be about a $1.5 billion dollar per year industry.

The cod fishery, moreover, was an employer of last resort.[269] Fishing was a dangerous and low-paying profession and earnings relied heavily on

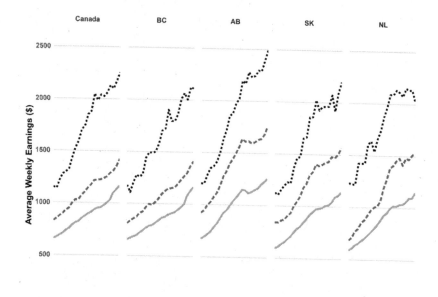

FIGURE 9 | Average weekly earnings by industry, 2001-22. Source: Statistics Canada Survey of Employment, Payrolls, and Hours. Author's graphic.

government support well before the collapse. In 1988, four years before the moratorium, the average fisher in Newfoundland and Labrador earned $11,000 less in net income than the average Newfoundland worker and more than one-third of fishers' income came from unemployment insurance.[270] Fishers could work for ten weeks and collect unemployment insurance for the balance of the year, a regime known colloquially as *Lotto 10/42*. The oil and gas industry is far from an employer of last resort; its workers are more diverse in skills, and the sector benefits from higher levels of education than was the case for the cod fishery. The payroll data in Figure 9 show that the average Alberta oil and gas worker earned more than double the average wage of workers in Alberta and almost three times the average Canadian worker's earnings in 2022.[271] All else being equal, an average Canadian job would have been a financial windfall for a Newfoundland fisher in the 1990s. Today, an average Canadian job would mean a substantial pay cut for the average oil and gas worker.

The transition plans implemented to cushion the impact of the cod fishery in Newfoundland achieved middling results, at best. They were late, ineffec-tive, and may have made some aspects of the transition more challenging.

Governments were hamstrung by the hope that a transition would not be required.[272] When transition became the only possibility, the only successful economic transition was to a different fishery, not to a different industry.[273] The cod moratorium offers little to inform our response to the energy transition—a markedly different challenge.

*   *   *

The Alberta coal phase-out is another comparison frequently made in discussions of just transition policies. Alberta's coal phase-out began with the establishment of a government-directed, orderly closure schedule, which appeals to some commentators. For example, in a TED Talk on transition planning, Notre Dame professor Emily Grubert recommends that governments set clear deadlines with ample notice. Such notice—Grubert argues for a decade—would "give communities enough time to create and implement plans," while a clear, legislated shut-down date "gives people enough confidence to commit to what can be an intense process."[274]

Alberta's government had a plan. But just as Prussian Field Marshal Helmuth von Moltke wrote that no battle plan survives first contact with the enemy, the Alberta coal phase-out plan was obsolete shortly after its first contact with market forces.[275]

In 2015, Alberta's six coal-fired generating stations served 64 percent of Alberta's net electricity load. Each relied mostly on coal from nearby mines. Coal did not appear to be a sunset industry: the province's newest coal plant, Keephills 3, had only come online in 2011. Keephills 3 and the Genessee 3 facility commissioned six years earlier were the two newest and most efficient coal units in Canada. About 2,000 workers, mostly in the mining sector, were directly employed in the provision of coal-fired electricity in Alberta.[276]

Rachel Notley was elected premier of Alberta in 2015, after having promised to phase out coal-fired power as part of an aggressive climate change policy platform. In November of that year, her government announced the Climate Leadership Plan, which committed to the aforementioned coal power phase-out by 2030. The policy was the product of extensive consultation with industry and stakeholders, which took place around the same time as the open houses at which I met my unforgettable miner. There was a lot of pressure from industry and stakeholders to maintain something close to the status quo in the electricity sector. Major players in the market and, to be honest, most of the technocrats I dealt with in government and in the province's grid operations during my time as chair of the Climate Change Advisory Panel felt that a phase-out of coal power by 2030 was not only infeasible but undesirable.[277] The government, based on the recommendations of my panel, forged ahead.

After announcing the coal phase-out, Alberta's government worked to solidify the type of plan recommended by just transition advocates: one that would bring key stakeholders, including electricity generators and organized labour, onside while providing certainty for workers and communities. In 2016, aided by a US power industry veteran acting as facilitator, the government signed a $1.1 billion compensation agreement with the owners of all coal power plants that had expected to operate beyond 2030.[278] This agreement served multiple purposes: first, in compensating generators for future revenues they would have earned, it turned opponents of the phase-out into supporters; second, it allowed the government to plan a staged closure of the coal units rather than risk the consequences of multiple plants closing over a short time in 2030; and, finally, the terms of the deal pushed the incumbent generators to invest in new renewable generation in the province which promised economic development in some regions affected by the coal phase-out.

The government also relied on input from a coalition of labour organizations, including the Alberta Federation of Labour and the country's largest union, Unifor, and it appointed an advisory panel to help establish support programs for workers and communities.[279] Provincial and federal programs provided income support worth up to 75 percent of coal workers' previous weekly earnings, payable for up to forty-five weeks, for those searching for new jobs. There were programs to bridge to retirement for workers over fifty-three years of age. Governments also offered relocation expenses, re-training programs, tuition vouchers, and career counselling for workers and support programs for affected communities. Alberta did exactly what advocates recommend to ensure that no one is left behind in a transition.

No one was prepared for what eventually transpired. From a GHG emissions perspective, Alberta's coal phase-out is a resounding success. It is arguably the most significant emissions reduction policy in Canadian history. The province is on-pace to phase out coal completely in 2023, only eight years after the commitment was announced, as plants have either retired or converted to natural gas much earlier than expected. From a worker and community transition perspective though, everything happened much faster than anyone would have or could have predicted. The government could force plants to shut down but had not considered the need to ensure that plants remained open long enough for workers and communities to adapt. Like Hemingway wrote of bankruptcy, the transition happened slowly at first, then all at once.

The transition happened so quickly that governments were unable to respond effectively. The federal government, for example, received the final report from its Task Force on Just Transition for Canadian Coal Power Workers and Communities after the Alberta coal phase-out was well underway.[280] The Auditor General reported in 2022 that the chaotic federal government response was plagued by unclear responsibilities and by government departments failing to secure protection for workers' pensions, or to deploy programs as mandated. The departments took what was termed a business-as-usual approach, adapting existing programs to attempt to buffer the coal transition. The Auditor General found that, by 2022, only four out of ten recommendations of the government's own task force had been implemented.[281] By then, only two coal units remained in operation in Alberta, with the balance having either shutdown or converted to natural gas. The list of the federal government's failings goes on and on. And the same government is now making promises to the oil and gas sector.

Alberta's response was more comprehensive, as noted above, but coal workers and communities still suffered. In Hanna, home of the Sheerness plants where the miner I spoke to at our open house worked, the town struggled to attract new investment.[282] As Gil McGowan, president of the Alberta Federation of Labour, stated in testimony to the House of Commons Standing Committee on Natural Resources, Alberta's just transition plan "was no panacea."[283] There are lessons to be learned from the coal transition in Alberta, but, as with the cod fishery, many do not readily apply to the transition facing the oil and gas sector. As McGowan testified, "it cannot and should not be used as a template for the energy transition that's already unfolding in oil and gas."

First and foremost, the size, scope, and scale of oil and gas differ from the Alberta coal industry. As McGowan testified, "there were only about 2,000 workers in Alberta affected by the coal phase-out," while the Alberta oil and gas sector alone employs well over 100,000. Furthermore, almost all the workers in the coal-fired power sector were union members. This provided the opportunity for engagement and planning that does not exist at the same scale in the oil and gas industry. The lack of union penetration also means that workers displaced from one oil and gas facility are unlikely to have any special preference given to them as they search for work at other facilities.[284] Even if we wanted to, McGowan concludes, "we cannot simply cut and paste what we did in the coal-fired power industry and apply it to oil and gas."[285]

McGowan argues that we need a much larger scale, economy-wide approach. This is largely where labour economist Jim Stanford lands, too.[286] While it might be desirable to reform our national social safety net, this generally is not what people have in mind when they talk about a just transition. Yet it might be exactly what we need if we are to deliver on the no-one-left-behind promise that governments are making to oil and gas workers.[287]

* * *

What about the coal phase-out in Germany? Is that a better point of comparison to what Canada's oil and gas industry is facing? You might expect that Germany would be committed to a rapid coal phase-out, given its declining coal production, its support for renewable electricity, and its aggressive stance on climate change. You would be wrong. Legislation passed in Germany in 2020 committed to a coal phase-out by 2038.[288]

The German coal phase-out has been a long time coming. German hard coal production peaked in 1957, and total German coal production declined by more than 50 percent after reunification, and has been declining slowly ever since. At its peak, over 600,000 people were employed in the German coal industry, but rapid declines reduced employment to fewer than 300,000 people before reunification and by another 100,000 or more in the first ten years after reunification.[289] German hard coal production ended in 2018, with the country now fully dependent on imports. German lignite coal production declined by more than half between 1989 and 2000, but some production remains today.[290] Coal consumption for electricity, provided in part by imports, has declined slowly but steadily since the mid-1990s, although the Russian invasion of Ukraine caused a rebound in coal power production and a sharp increase in coal imports in 2022.[291]

The German coal transition is not a story of a prosperous industry facing a new threat from climate policy. The German coal transition, if one can call a sixty-year industrial history a transition, is a story of an economically unviable sector, undermined by global competition, but kept afloat at great expense through state intervention. The German coal transition began when trade agreements led to the ending of price controls in 1956.[292] In the same way as Canada's manufacturing downturn was buffered by its resource industry, the early years of the coal collapse did not lead to massive unemployment in Germany's principal coal-producing region, the Ruhr Valley. On the contrary, "Germany's economic miracle during the postwar period was such that displaced workers had ample opportunity in booming steel and metal sectors."[293] Substantial government support programs for miners and mining communities are a long-standing German tradition.[294] At the time of a 2007 agreement to end mining, for example, the German coal industry was receiving $3.2 billion per year in subsidies which, when averaged over mine workers, worked out to more than the average miner would have earned in a year.[295]

There is a plausible future in which the oil sands of Alberta find themselves in a situation similar to that of the German coal industry. But that is not where we are today. The German government has been using policy to keep the coal industry afloat for decades. There is no meaningful comparison between that and a policy-forced transition away from oil and gas in Canada. The latest German coal transition allows the government to reduce the economic drain on its coffers and to direct resources directly to workers rather than cycling it through a subsidized mining industry. Canadian governments will have to contemplate the replacement of revenues, wages, taxes, and royalties from a profitable industry in the event of a transition away from oil and gas. There is no comparison to be had.

* * *

What about the Ontario coal phase-out? It, too, provides few useful lessons for prospective transition plans for fossil fuels.

The end of coal in Ontario was initially contemplated as early as 1986 when Ontario Hydro began planning to mothball its coal plants in response to policies to control acid rain.[296] These plans were altered by changes in Ontario's electricity market, including capacity shortfalls owing to the province's nuclear refurbishment.[297] With nuclear capacity reduced, increased coal generation coincided with the emergence of smog as a serious concern in Southern Ontario. Concerns about health issues with coal generation were the primary drivers of an all-party consensus in the 2003 election on the need to phase out coal and an eventual recommendation in an administrative review of the power system to complete a coal phase-out by 2015.[298]

The Ontario coal power industry had little in common with today's oil and gas industry and not even much in common with the Alberta coal power sector. The size of the Ontario coal power industry was similar to that in Alberta, in that fewer than 2,000 workers were directly employed in coal power before the phase-out began, although these were almost all employed by the same Crown corporation, which was not the case in Alberta.[299] Both the Ontarian and Albertan phase-outs were enabled by ready alternatives, mostly natural gas. Some coal plants in both jurisdictions were converted to burn other fuels. The similarities end there. Ontario's five coal plants were relatively old and inefficient and most of the coal was imported, so there were fewer jobs tied to coal power generation than was the case because of on-site mining in Alberta. Health concerns meant there was pan-partisan and technocratic support for a phase-out, which was also not the case for Alberta coal and is certainly not the case with phasing out oil and gas production.[300] The Ontario coal phase-out offers few useful lessons. Any potential transition away from fossil fuels in Canada would be orders of magnitude greater in scale and scope.

* * *

On April 30, 2023, as I was preparing this volume, the CBC's *What On Earth* radio program aired an episode on the just transition entitled "How to Ditch Fossil Fuels Without Leaving Workers Behind."[301] A nice thought, to be sure, and one that piqued my interest. The show featured, as usual, a conversation with a German coal miner who raved about the transition programs he had experienced. When his job site shut down, he was bused to another site operated by the same state-controlled company. Some of his colleagues had been retrained, in physiotherapy, for example, but the featured miner had retired early and was content to watch his favourite football team. A nice life.

It was not the miner's story that really caught my attention in that episode. It was a segment featuring Angela Carter, Associate Professor at the University of Waterloo and an energy transitions specialist with the International Institute for Sustainable Development. She spoke of Denmark's oil and gas phase-out and its transition to wind turbines. Denmark's transition often features in studies of fossil fuel phase-outs, not least because the symbolism of a wind-turbine industry taking the place of the oil and gas industry is powerful. It is also a terrible comparison to Canada's oil sector.

Why?

Denmark's oil and gas sector has long been in decline and is a fraction of the size of Canada's. Danish oil production and exports peaked in the mid-2000s and have been in decline since, with oil production today at less than 20 percent of its peak. Danish gas production is today less than a third of what it was at its mid-2000s peak and it is still falling. In Canada, oil and gas production continues to break records. Outlooks for oil production, even under more stringent climate policies, see stable production and exports at least through the end of this decade.[302] A Canadian Climate Institute study estimated that the Danish oil and gas phase-out policy would cost $2.5 billion in foregone government revenue between 2020 and 2050.[303] By comparison, the direct government revenues (royalties and lease payments, not corporate or personal income taxes) from oil and gas extraction to the Government of Alberta add up to $2.5 billion every two months or so.

The *What on Earth* podcast asked, "what would motivate a major producing state to phase out oil production? And what could we learn from them?"

Denmark is not a major oil and gas producing state in the way Canada is today. I think the biggest thing we can learn from these false comparisons is that we need to compare industries in similar circumstances rather than look for easy answers. The reality, in Denmark, was one of a relatively small oil and gas industry in sharp decline with expectations for a continued decline to come. That is not remotely comparable to the prospect of a policy-forced transition away from oil and gas in Canada.

* * *

The just transition narrative that derives from examples like the seemingly never-ending German coal phase-out or the revered wind-turbine transition in Denmark is one that flows from a specific set of circumstances: an industry in secular decline, with a comparative disadvantage in local or global markets; public ownership or substantial public involvement; small, concentrated, highly-unionized, and easily identifiable labour force; and limited foregone revenues from the transition or, in some cases, savings to government coffers. In many cases, income and/or price supports have been a long-term part of the industry. None of these is present in the Canadian oil and gas sector today.

We are told that the government, to enable a just transition, needs a clear plan for phasing out the industry, a plan to transition the workers toward retirement and a solid social safety net to cushion the fall for younger workers. It is hard to imagine what that might look like in the Canadian oil and gas sector.

In his paper on a just transition for the Canadian oil and gas sector, Jim Stanford provides few details, but states that Canada's 2050 net-zero target needs to be "reinforced with more specific timetables for the phase-out of various components of the overall energy system."[304] Similar notions of a planned transition are echoed in work by Professors Emily Grubert and Sara Hastings-Simon, although their focus is more on planning to ensure the presence of key elements of both a viable, shrinking fossil-fueled energy system and a growing clean energy system.[305] For example, they worry more about who will produce, process, and transport the small amounts of natural gas that an electricity grid operator might assume will remain available for power generation. I confess that I cannot visualize how such a planned phase-out might work even with a relatively concentrated industry like the oil sands, let alone the entirety of the oil and gas sector.

How is the federal or even the provincial government going to send a clear signal to oil and gas companies about the specific timetable for their projects' phase-out? You could set a timetable in a spreadsheet with relative ease. But who decides which facilities shut down when? How would it be enforced? Would firms treat the signal as credible? Would workers continue to work at a facility marked for an early shutdown? And would only workers displaced according to the plan be eligible for special social programs and other preferential treatment? It sounds simple to say we should have a plan, but it would be no simple task to formulate and implement one.

Complexities notwithstanding, let us suppose governments were to set a timeline for a particular oil sands mine to shut down in 2030. This hypothetical scenario raises another set of questions. How would the facility remain a going concern through the voluntary exits of its best workers? And how would the facility renew its operating workforce given a looming forced shutdown? It likely would not be able to do so. It is also not clear how government would ensure that this hypothetical mine's operators undertake much-needed reclamation work, after having lost much of the previously expected value of future production from their project. As soon as a binding end-date is set for a facility, a cascading set of challenges is likely to arise and lead to one of two unsatisfactory endings: an early, chaotic shutdown, or a government takeover. Hundreds of oil sands sites and many hundreds of smaller facilities would face similar predicaments.

The questions above address only the facilities themselves, not the communities in which they operate and in which their workers live. Stanford, like many others, recommends that a government plan support for regional adjustment and diversification. He is confident that other industries can replace the oil and gas sector in many regions of the country. I have spent a lot of time in regions that could use the equivalent an oil sands industry right now. We do not have to wait for the oil sands to shut down to create these new industries that Stanford and others promise could fill a future gap. If all you need to spur new development is a town that has lost its key employer, we have got hundreds of those available right now. If new industries paying high wages can be created by government fiat, incomes would be much higher than they are today many regions of the country.

The reality is that governments cannot choose comparative advantage nor can they will into existence a lucrative industry. We can create jobs with government work and subsidized diversification, but we cannot recreate something of the size and scale of the oil sands. Governments cannot make that much water run uphill.

The truth is that in an industry as decentralized as the Canadian oil and gas sector, there is not going to be a planned, orderly transition. And, like the Alberta coal phase-out, any attempts to plan one will not survive contact with market forces. For just one example of what could go wrong, consider that a lot of writing and commentary on a just transition away from fossil fuels assumes that oil and gas prices will decline smoothly and predictably in a climate-constrained world, as per IEA or similar projections. Even assuming a well-planned transition in Canada, a smooth oil price decline is by no means guaranteed. As mentioned earlier, oil prices might not drop and might even increase, or they might crash. Each of these outcomes would present challenges to any planned transition.

Oil prices, again, result from the interaction of global supply and demand, which both depend on investments and other decisions made in previous years. The oil market of the future will see one or more of four broad possibilities: high or low oil prices could arise with low consumption in a world acting on climate change; high or low oil prices could also arise with high consumption in a world failing to act. And if everyone expects low prices in the future, that could create the market conditions for high or, at least, higher prices. The prices we see are the product of equilibrium between supply and demand.

How does that affect planning for a so-called just transition as some envision it? Imagine that we find ourselves in 2030 and the world is facing substantial climate and geopolitical upheaval. Actions to reduce emissions are sporadic, uncoordinated, and targeted for convenience and political expediency rather than effectiveness. Global oil consumption remains high, or at least high relative to expectations. Global oil majors, under pressure from shareholders concerned about climate change, have scaled back investment in new production. Global oil supply is tight and prices are high.

Now, imagine that this is all happening at the time when, according to the transition plan, multiple oil sands projects that are profitable under the prices of the day are slated to close. Workers, seeing high oil prices and profits, might not expect governments to follow through in forcing the early shutdown of the facility. And with high oil prices driving high wages, no one would be looking to transition to an average job elsewhere until absolutely necessary. Workers would likely be right: even in the extreme case of asbestos mining in Quebec, it was global market forces and not government policy that finally shut down the last facility. Our governments fought to keep those mines open and even pushed to re-open them after they were closed.[306] Have we ever had a case of government locking the gates to a profitable major industrial project and sending the workers home for good? Certainly not in an industry as large and profitable as oil and gas is today. A forced shut-down of profitable oil sands facilities would be almost unimaginable if oil prices stay high.

Or prices could drop faster than expected, but this does not make a planned transition any easier. Global oil producers might bet too heavily on increasing global consumption, or electric vehicles could take over the market faster than we expect, pushing oil prices down faster than predicted even in today's scenarios for aggressive action on climate change. With oil prices tanking, the oil sands transition might start to look like the Alberta coal transition: companies with limited remaining project lives might not be willing to endure a period of low prices and losses, especially if aggressive climate policies have also reduced their potential future profits. The fixed-date shutdown plan makes it more likely that the facility will shut down early.

Furthermore, as discussed earlier, a period of low revenues combined with shortened project lives would (or, should) trigger demands for additional environmental reclamation security deposits. This cash call could, in turn, accelerate firms' shutdown decisions and leave a massive liability on the landscape. In these cases, as in the Alberta coal power phase-out, worker displacements would happen much earlier than planned.

Each of these scenarios leads to unsatisfactory outcomes and people left behind. And both are more likely than a smooth, planned transition with smoothly declining commodity prices that so many want to imagine.

* *.*

So what can we do? We need to think about this not like the phase-out of coal in Ontario where we can, at least hypothetically, plan for a smooth transition of workers in a government-owned enterprise, nor like the cod fishery where a biophysical constraint will determine most of the actions governments can take. The energy transition will likely look most like the economic evolution seen in Ontario in the early 2000s when manufacturing in areas where comparative advantage was lost, or where new subsidies were not forthcoming from government, faded slowly into a less relevant part of the economy. Or it might resemble the transition faced by the lumber industry in my family's home of Miramichi, in northeastern New Brunswick. It is going to be bumpy and unpredictable. Transition job losses will be difficult to differentiate from the usual churn of the economy. As a small, open economy, we are price-takers not price-setters in the global markets and our governments cannot expect to sustain large industries solely through domestic trade or price controls. Through subsidies, governments can pick winners and losers, but they would be unable to create comparative advantage that attracts waves of foreign and domestic investment that we have seen in the oil and gas sector.

When Ontario's manufacturing sector was shrinking, government did provide substantial support to two automobile manufacturers, but there was no plan to do so on a broader scale, nor were there calls for a planned phase-out of existing plants over time to get ahead of the disorderly transition that global markets would impose. There was no hubris that governments could save the industry and all of its workers from the pain of a transition, or not like we're seeing with regard to oil and gas, anyway. There was no glossy plan to shift to an Ontario with less manufacturing and no election campaign platforms premised on an orderly shutdown with no one left behind. Nothing of the sort. And there was certainly no one telling the workers fortunate enough to remain employed that they'd be better off without the jobs they still had because the factories in which they worked faced significant future market risk.

Most of the transition in Ontario was buffered by growth in other sectors, both within the province and across the country. And, where it was needed, we relied on our national and provincial social safety nets to support those workers who did lose their jobs. And we will do so again.[307]

That will be unsatisfying to many, but the oil and gas industry does not lend itself to a planned phase-out and to promise such an orderly and painless transition is a false guarantee. We should not pretend that we can tell our young people which companies in the oil and gas sector will survive, which projects will still be operating in twenty years, or which jobs will end in an early termination. We simply cannot know and we should not pretend we can know.

We should and I hope we will continue to educate our young people on the realities of climate change, on the likely macroeconomic implications of global action to reduce emissions, and on the risks inherent in all sectors of the economy. But we must also accept that for some students graduating today, their best job options are in the oil and gas sector. That is okay. And, just as people may bet on the wrong technology or find that their first choice of career is not for them, we should make sure we have a system that helps them avoid being left behind permanently.

Technology and policy have and will continue to drive economic transitions. There will be people left behind. Governments cannot and should not promise otherwise. I often think of the miner that I met at our climate panel open house and wonder what has happened to him since his coal mine closed. In addition to that miner, I think of all the other Albertans who, due to the oil price downturn or policy changes, found themselves out of work long before the Sheerness mine closure. I expect we will see many more worker transitions for which we cannot plan than those for which we can engineer an orderly phase-out. Let us make sure we have a system that works for those people, and not pretend that we can do things we cannot.

8 | Conclusions and omissions

Climate change, again, is the environmental, political, and societal challenge of our time. The goal of this volume was to highlight the ways in which we, as Canadians, have not always responded constructively to the challenge it presents, and to push back against some of the half-truths and clever excuses that have come to define climate change policy debates in Canada.

We can adapt to some impacts of climate change, but we should not forgo low-cost opportunities to mitigate emissions either, or we will lock-in more suffering than we should. We are a cold country, but we will still be affected in many catastrophic ways by climate change, and these effects will only be limited by global action. To that end, while we contribute less than 2 percent of global emissions, we have caused more than our share of the problem, and we have a responsibility and a strategic imperative to be full participants in a collective response. We will not be able to argue for global action while refusing to take such action ourselves.

The world will still use oil and gas for decades to come, but that provides little assurance that our fossil fuel industries will not be affected by action on climate change. Oil and gas demand is on a collision course with action on climate change, and the more we look to hide behind solutions like carbon capture and storage, the more likely we are to fail to plan for the much larger challenges we face in the coming transition. Many are, indeed, too quick to dismiss the potential resilience of our oil sands industry to action on climate change. Perhaps this is, in part, because they have been conditioned by industry communications to think of the oil sands industry as more precarious than it is.

And, speaking of the energy transition, while wind and solar power are revolutionizing global energy systems, they do face important limits in a northern country like Canada. And, these limits demand research focused on Canada, not research adapted to Canada but anchored in the realities of our southern neighbours. And, finally, while some are quick to imagine that government action can create new industries to replace lost jobs, wages, and tax revenues, and provide a *just transition* away from fossil fuels, the reality is that the coming energy transition will be unprecedented, painful, and costly to many here in Canada even as it will also have substantial global benefits.

I believe that Canada is up for each of the challenges I have raised, but that we will be more prepared for what climate change will bring if we confront it head-on rather than with excuses and easy soundbites.

One challenge with a project like this is that there are more little lies, false promises, and easy soundbites that I have left untouched than the six that I have examined in detail. As I wrap up, I will identify a few more.

I did not spend much time dissecting global climate change mitigation goals, and I mostly took the 1.5°C and 2°C goals in the Paris Agreement as given. I did consider including a chapter on climate thresholds and the mentality that has emerged in some quarters that all would be lost if global climate change were to pass either of these *magic numbers*. That is not the case. While the 1.5°C and 2°C thresholds were chosen to provide concrete goals to rally action on climate change, things do not stop getting worse the further we go past these targets. On the contrary, things get a lot worse. From an economics perspective, the social cost of emissions is increasing the more emissions increase, so we should be willing to do more to reduce emissions today if we expect to cross these thresholds. But, as an economist, I felt I was best to leave a detailed study of the physical effects of climate change and the importance of specific thresholds like the 1.5°C and 2°C goals to authors like Katharine Hayhoe and her book, *Saving Us.*

I looked at statements that the world will continue to use (enough) oil and gas, but largely ignored the potential for Canadian LNG in a carbon-constrained world. LNG is challenging because, if you look at it in a vacuum, substituting Asian coal combustion with Canadian natural gas liquefied and shipped by tanker from BC leads to a material reduction in emissions. But, as cleaner sources of energy become cheaper, there are other, lower emissions alternatives for coal. And most analyses show that we already have enough global LNG capacity to serve likely demand in a carbon-constrained world. This does not mean that potential Canadian LNG plants would become stranded, but it is a possibility as Canadian LNG is more expensive than some global competition. On the whole, an exploration of the business case for LNG would have been an interesting addition to this volume, given more time and space.

Another common refrain that I considered addressing was that resources are an exclusive provincial jurisdiction under the Canadian Constitution. This is more of an economics than a law text, but Canadians could benefit from a better understanding of the Canadian division of powers, so leaving this subject aside was difficult. It is trite law that, per section 92A of the *Constitution*

Act, 1867, provinces have the exclusive jurisdiction to legislate in relation to most aspects of natural resource development and electricity generation.[308] Provinces also hold proprietary rights over Crown-owned resources other than those on federal lands. Neither serves to confer immunity from the effects of valid federal legislation, whether that legislation is in the form of taxes, environmental regulations, fisheries rules, export tariffs, or rules made with respect to cross-border infrastructure such as pipelines. In the coming years, there will be a series of cases, beginning with the forthcoming Supreme Court decision on the Impact Assessment Act (Bill C-69), which may redefine federal incursion into provincial jurisdiction over resources and electricity, but the core elements of what I have written seem unlikely to change. There is no inalienable provincial right to the development of natural resources where such development might contravene federal legislation, and no blanket restriction on federal legislation that prevents it from affecting provincial resource production. I would have enjoyed including a deeper exploration of these issues in this volume as well.

Some readers will be surprised that I have written little about carbon taxes or carbon pricing in general. I remain a strong supporter of Canada's carbon pricing legislation, and there are many little lies and half-truths about carbon pricing that I could have addressed, including from those who argue that carbon taxes do not lead to lower emissions than would otherwise be the case or those who claim that consumer rebates eliminate the incentive to reduce emissions. But the subject I most wanted to include highlights an error I (and others) have made in the past in discussing whether lower-income Canadians are better off with carbon pricing. Economists, myself included, sometimes forget—or at least forget to mention—that there is a distribution of impacts of carbon pricing within income deciles or quartiles. For example, when we emphasize the average impact of carbon pricing on those in the bottom 30 percent of the income distribution, we should always acknowledge that not everyone is the same within that group. While it is true on average, it is not always the case that high-income households "bear a larger cost burden compared to lower-income households," nor is it always the case that a lower-income household's spending is "less carbon intensive," as stated in a recent report from the Office of the Parliamentary Budget Officer.[309] There will be households within each group that have very different outcomes from the average, and certainly some lower-income households will not be fully compensated by the rebates that accompany the carbon price. I would have loved to have had more room to consider these and other carbon-tax-related issues in this collection.

I also considered including a challenge to the claim seen in some circles that electric vehicles are no better for the environment than internal combustion engines unless the power grid is very clean. This is almost always not the case because of the efficiency of electric motors. Electric vehicles travel up to five times further per unit energy than gasoline vehicles, mostly because so much of the energy used in an internal combustion engine is wasted as heat rather than used to power the car.[310] There will be efficiency losses between the power plant and the vehicle, and these will be larger with a fossil-intensive grid, but the efficiency difference is large enough to overcome almost all of that. For example, the IEA found that electric vehicles have lower life-cycle emissions than an internal combustion engine vehicle even if the electric vehicles are charged on power grids with an average emissions intensity equal to that of Alberta's newest coal plants.[311] These results do vary depending on the production emissions associated with the electric vehicle battery and the life cycle emissions of the gasoline used to power the conventional engine, and so it is possible to construct a scenario in which an internal combustion engine is a lower-emissions alternative in a very emissions-intensive power grid.[312] But as global clean energy supply ramps up, emissions-intensive power grids will soon be a thing of the past.[313] In most locations worldwide, and most everywhere in Canada, you can be confident that your electric vehicle will have lower life-cycle emissions than a comparable internal combustion engine vehicle.

The last half-truth that I did not take the time to address is a subtle one, but one we hear often: "this is not a climate change policy." There is a temptation to classify only those government actions explicitly designed to reduce emissions as climate change policies. That is wrong. Governments engage in climate change policy every time they encourage new economic activity, authorize infrastructure spending, create zoning laws or building codes, and sign and implement trade agreements. Some of these actions increase emissions above where they might otherwise be and, in some cases, governments are explicit about that.[314] Other actions may increase emissions quietly, while others serve to reduce emissions. Some policies, like climate risk disclosure policies for securities, do not force changes in emissions, but we have seen how shareholder pressure has driven major emissions reduction and divestment decisions motivated by climate change.[315] In short, (almost) every policy is climate policy. The federal government has been working for some time on expanding a so-called *climate lens* to many more of its decisions, and that is a great start. Governments need to take account of climate change mitigation and adaptation in a lot more of the decisions they make, from infrastructure to housing to fiscal policy.

I hope that you will be on the lookout for these and other little lies and half-truths as we navigate the coming energy transition and confront the hard truths of climate change.

Notes

1 Rebecca Ananda, "'Tell Your Parents You're at the Library': A Brief History of Ryerson's Imperial Pub," *Ryersonian* (blog), April 2014, https://ryersonian.ca/tell-your-parents-youre-at-the-library/.

2 Katharine Hayhoe, *Saving Us: A Climate Scientist's Case for Hope and Healing in a Divided World* (Simon and Schuster, 2021), 49.

3 Intergovernmental Panel on Climate Change, "AR6 Synthesis Report (Longer Report)," 2023, https://www.ipcc.ch/report/ar6/syr/; Intergovernmental Panel on Climate Change, "AR6 Synthesis Report (Summary for Policymakers)," 2023, https://www.ipcc.ch/report/ar6/syr/.

4 Intergovernmental Panel on Climate Change, "AR6 Synthesis Report SPM," 6; Intergovernmental Panel on Climate Change, "Summary for Policymakers," in *Climate Change 2021: The Physical Science Basis. Contribution of Working Group I to the Sixth Assessment Report of the Intergovernmental Panel on Climate Change [Masson-Delmotte, V., P. Zhai, A. Pirani, S.L. Connors, C. Péan, S. Berger, N. Caud, Y. Chen, L. Goldfarb, M.I. Gomis, M. Huang, K. Leitzell, E. Lonnoy, J.B.R. Matthews, T.K. Maycock, T. Waterfield, O. Yelekçi, R. Yu, and B. Zhou (Eds.)]* (Cambridge, UK: Cambridge University Press, 2021), 4, doi:10.1017/9781009157896.

5 Intergovernmental Panel on Climate Change, "AR6 Synthesis Report SPM," 6; Intergovernmental Panel on Climate Change, "AR6 Working Group 1 Summary for Policymakers," 4.

6 Intergovernmental Panel on Climate Change, "AR6 Working Group 1 Summary for Policymakers," 4. The list of false climate change hypotheses is taken from Hayhoe, *Saving Us*, 56–58.

7 Intergovernmental Panel on Climate Change, "AR6 Working Group 1 Summary for Policymakers," 5.

8 Intergovernmental Panel on Climate Change, "AR6 Synthesis Longer Report". See also International Energy Agency, "CO2 Emissions in 2022," 2023, https://www.iea.org/reports/co2-emissions-in-2022.

9 Intergovernmental Panel on Climate Change, "AR6 Synthesis Longer Report," 17.

10 Hayhoe, *Saving Us*.

11 "The Paris Agreement, United Nations Framework Convention on Climate Change," 12 December 2015, UNTS No 54113 (entered into force 4 November 2016) § (2015).

12 United Nations Environment Programme, "Emissions Gap Report 2022: The Closing Window," October 2022, XX, http://www.unep.org/resources/emissions-gap-report-2022.

13 Justin Ritchie and Hadi Dowlatabadi, "Why Do Climate Change Scenarios Return to Coal?," *Energy* 140 (December 2017): 1276–1291, doi:10.1016/j.energy.2017.08.083;

Zeke Hausfather and Glen P. Peters, "Emissions–the 'Business as Usual' Story Is Misleading," *Nature* 577, no. 7792 (January 2020): 618–620, doi:10.1038/d41586-020-00177-3; Dirk-Jan van de Ven et al., "A Multimodel Analysis of Post-Glasgow Climate Targets and Feasibility Challenges," *Nature Climate Change*, May 2023, 1–9, doi:10.1038/s41558-023-01661-0.

14 Malte Meinshausen et al., "Realization of Paris Agreement Pledges May Limit Warming Just below 2 °C," *Nature* 604, no. 7905 (April 2022): 304–309, doi:10.1038/s41586-022-04553-z; van de Ven et al., "A Multimodel Analysis of Post-Glasgow Climate Targets and Feasibility Challenges."

15 Paul A. Samuelson, "The Pure Theory of Public Expenditure," *The Review of Economics and Statistics* 36, no. 4 (1954): 387–389, doi:10.2307/1925895.

16 International Energy Agency, "Oil 2023," 2023, https://www.iea.org/reports/oil-2023.

17 van de Ven et al., "A Multimodel Analysis of Post-Glasgow Climate Targets and Feasibility Challenges". See also Intergovernmental Panel on Climate Change, "AR6 Synthesis Report SPM," 11.

18 Author's calculations based on four-model average annual emissions decline rates by decade from the long term targets scenario in van de Ven et al., "A Multimodel Analysis of Post-Glasgow Climate Targets and Feasibility Challenges."

19 Government of Canada, "Canada's 2021 Nationally Determined Contribution (NDC)," UNFCCC, 2021, https://www4.unfccc.int/sites/ndcstaging/PublishedDocuments/Canada%20First/Canada%27s%20Enhanced%20NDC%20Submission1_FINAL%20EN.pdf.

20 John Holdren, "Remarks on Climate Change to the National Academies of Science," November 2009, https://www.c-span.org/video/?290231-1/john-holdren-remarks-climate-change.

21 I have seen Holdren quoted in many sources over the years, but, in this instance, I was reminded of it from a quote in Hayhoe, *Saving Us*, 125.

22 Bjorn Lomborg, "Climate Change Calls for Adaptation, Not Panic," Wall Street Journal, October 2021, https://archive.md/SfEGz.

23 Economists define a *public good* as one for which there is non-rivalry (my consumption does not crowd out your opportunity to consume) and non-excludability (I can't prevent you from accessing the good) in consumption.

24 Intergovernmental Panel on Climate Change, "AR6 Synthesis Report SPM," 23.

25 van de Ven et al., "A Multimodel Analysis of Post-Glasgow Climate Targets and Feasibility Challenges."

26 Intergovernmental Panel on Climate Change, "AR6 Synthesis Longer Report," 11.

27 Intergovernmental Panel on Climate Change, "AR6 Synthesis Report SPM," 16. The balance of this paragraph is drawn from the same source, and so references are omitted.

28 Intergovernmental Panel on Climate Change, "AR6 Working Group 1 Summary for Policymakers," 15.

29 Kevin Rennert et al., "Comprehensive Evidence Implies a Higher Social Cost of CO2," *Nature* 610, no. 7933 (October 2022): 687–692, doi:10.1038/s41586-022-05224-9; Richard S. J. Tol, "Social Cost of Carbon Estimates Have Increased over Time," *Nature Climate Change*, May 2023, 1–5, doi:10.1038/s41558-023-01680-x.

30 R H Coase, "The Problem of Social Cost," *The Journal of Law and Economics* 3 (1960): 44.

31 Rennert et al., "Comprehensive Evidence Implies a Higher Social Cost of CO2."

32 See estimates for the United States in the United States Environmental Protection Agency, "Supplementary Material for the Regulatory Impact Analysis for the Supplemental Proposed Rulemaking, 'Standards of Performance for New, Reconstructed, and Modified Sources and Emissions Guidelines for Existing Sources: Oil and Natural Gas Sector Climate Review': EPA External Review Draft of Report on the Social Cost of Greenhouse Gases: Estimates Incorporating Recent Scientific Advances," 2022, https://www.epa.gov/system/files/documents/2022-11/epa_scghg_report_draft_0.pdf, and estimates adapted for Canada in Environment and Climate Change Canada, "Social Cost of Greenhouse Gas Emissions," April 2023, https://www.canada.ca/en/environment-climate-change/services/climate-change/science-research-data/social-cost-ghg.html. Previous Canadian estimates are avaialble in Environment and Climate Change Canada, "Technical Update to Environment and Climate Change Canada's Social Cost of Greenhouse Gas Estimates," Environment and Climate Change Canada, March 2016, https://perma.cc/G2G2-JHXE.

33 Rennert et al., "Comprehensive Evidence Implies a Higher Social Cost of CO2," 690.

34 Rennert et al., 690.

35 Rennert et al., 690.

36 United States Environmental Protection Agency, "Supplementary Material for the Regulatory Impact Analysis for the Supplemental Proposed Rulemaking, 'Standards of Performance for New, Reconstructed, and Modified Sources and Emissions Guidelines for Existing Sources: Oil and Natural Gas Sector Climate Review': EPA External Review Draft of Report on the Social Cost of Greenhouse Gases: Estimates Incorporating Recent Scientific Advances," tbl. 3.2.1.

37 James Hansen, "Silence Is Deadly," June 2011, http://www.columbia.edu/~jeh1/mailings/2011/20110603_SilenceIsDeadly.pdf; Bill McKibben, "Barack Obama, the Carbon President," The Guardian, June 2011, https://www.theguardian.com/commentisfree/cifamerica/2011/jun/03/barack-obama-carbon-president.

38 Andrew Leach, "On the Potential for Oil Sands to Add 200ppm of CO2 to the Atmosphere," Rescuing the Frog (blog), June 2011, http://andrewleach.ca/?p=562. For a write-up of the fight over oil sands emissions impacts, see Chris Turner, The Patch (Simon and Schuster, 2017), https://www.simonandschuster.ca/books/The-Patch/Chris-Turner/9781501115103, Chapter 11 (p. 254).

39 Lomborg, "Climate Change Calls for Adaptation, Not Panic."

40 International Energy Agency, "Curtailing Methane Emissions from Fossil Fuel Operations: Pathways to a 75% Cut by 2030," 2021, https://www.iea.org/data-and-statistics/charts/global-methane-abatement-cost-curve-by-policy-option.

41 For example, in International Energy Agency, "World Energy Outlook 2022," October 2022, 229, https://www.iea.org/reports/world-energy-outlook-2022, the IEA documents over US$500 billion in direct fossil fuel consumption subsidies. In their 2022 analysis, International Energy Agency, "Fossil Fuels Consumption Subsidies 2022," 2022, https://www.iea.org/reports/fossil-fuels-consumption-subsidies-2022, these figures are updated to over $1 trillion worth of consumption subsidies in the wake of the Russian invasion of Ukraine.

42 Hayhoe, Saving Us, 126.

43 Joe Oliver, "Here's a Truth Few Dare to Utter: Canada Will Benefit from Climate Change," Financial Post, August 2019, https://financialpost.com/opinion/joe-oliver-heres-a-truth-few-dare-to-utter-canada-will-benefit-from-climate-change.

44 Chris Lafakis et al., "The Economic Implications of Climate Change," Moody's

Analytics, June 2019, https://www.moodysanalytics.com/-/media/article/2019/economic-implications-of-climate-change.pdf.

45 Katharine Ricke et al., "Country-Level Social Cost of Carbon," *Nature Climate Change* 8, no. 10 (October 2018): 895–900, doi:10.1038/s41558-018-0282-y.

46 Lee Hannah et al., "The Environmental Consequences of Climate-Driven Agricultural Frontiers," *PLOS ONE* 15, no. 2 (February 2020): e0228305, doi:10.1371/journal.pone.0228305.

47 Roberto Roson and Martina Sartori, "Estimation of Climate Change Damage Functions for 140 Regions in the GTAP 9 Data Base," *Journal of Global Economic Analysis* 1, no. 2 (December 2016): 78–115, doi:10.21642/JGEA.010202AF.

48 For other examples of articles that estimate potential benefits of climate change in Canada, see Michelle J. Reinsborough, "A Ricardian Model of Climate Change in Canada," *The Canadian Journal of Economics / Revue Canadienne d'Economique* 36, no. 1 (2003): 21–40, Robert Mendelsohn and Michelle Reinsborough, "A Ricardian Analysis of US and Canadian Farmland," *Climatic Change* 81, no. 1 (March 2007): 9–17, doi:10.1007/s10584-006-9138-y, Marian Weber and Grant Hauer, "A Regional Analysis of Climate Change Impacts on Canadian Agriculture," *Canadian Public Policy* 29, no. 2 (2003): 163–180, doi:10.2307/3552453, or Afshin Amiraslany, "The Impact of Climate Change on Canadian Agriculture : A Ricardian Approach," April 2010, https://harvest.usask.ca/handle/10388/etd-05252010-102012.

49 Matthew E. Kahn et al., "Long-Term Macroeconomic Effects of Climate Change: A Cross-Country Analysis," *Energy Economics* 104 (December 2021): 4, doi:10.1016/j.eneco.2021.105624.

50 See, for example, Marshall Burke, Solomon M. Hsiang, and Edward Miguel, "Global Non-Linear Effect of Temperature on Economic Production," *Nature* 527, no. 7577 (November 2015): 235–239, doi:10.1038/nature15725, Riccardo Colacito, Bridget Hoffmann, and Toan Phan, "Temperature and Growth: A Panel Analysis of the United States," *Journal of Money, Credit and Banking* 51, no. 2–3 (2019): 313–368, doi:10.1111/jmcb.12574, or Kahn et al., "Long-Term Macroeconomic Effects of Climate Change". The methodology used in the latter two papers was adopted for Canada in Philip Bagnoli et al., "Global Greenhouse Gas Emissions and Canadian GDP" (Office of the Parliamentary Budget Officer, November 2022), https://www.pbo-dpb.ca/en/publications/RP-2223-015-S--global-greenhouse-gas-emissions-canadian-gdp--emissions-mondiales-gaz-effet-serre-pib-canadien. See also a useful review and methodological analysis in Richard G. Newell, Brian C. Prest, and Steven E. Sexton, "The GDP-Temperature Relationship: Implications for Climate Change Damages," *Journal of Environmental Economics and Management* 108 (July 2021): 102445, doi:10.1016/j.jeem.2021.102445.

51 The estimated relationships in Burke, Hsiang, and Miguel, "Global Non-Linear Effect of Temperature on Economic Production," form the basis for the estimated country-level social costs of carbon in Ricke et al., "Country-Level Social Cost of Carbon," which is why that analysis shows that Canada would benefit from climate change.

52 Government of Canada, Department of Environment and Climate Change, "Temperature Change in Canada," August 2022, https://www.canada.ca/en/environment-climate-change/services/environmental-indicators/temperature-change.html.

53 Natural Resources Canada, "Climate Change Adaptation in Canada" (Natural Resources Canada, November 2022), https://natural-resources.canada.ca/climate-change/what-adaptation/10025.

54 Dylan G. Clark et al., "Due North," Canadian Climate Institute, 2022, https://climateinstitute.ca/reports/due-north-costs-of-climate-change/.

55 "What Is Permafrost?," NASA Climate Kids, accessed May 15, 2023, https://climatekids.nasa.gov/permafrost/.

56 Natural Resources Canada, "Geoscience: Climate Change" (Natural Resources Canada, October 2013), https://natural-resources.canada.ca/earth-sciences/earth-sciences-resources/geoscience-climate-change/10900.

57 R. Iestyn Woolway et al., "Lake Ice Will Be Less Safe for Recreation and Transportation Under Future Warming," *Earth's Future* 10, no. 10 (2022): e2022EF002907, doi:10.1029/2022EF002907, citing Danny Blair and David Sauchyn, "Winter Roads in Manitoba," in *The New Normal: The Canadian Prairies in a Changing Climate*, ed. David Sauchyn, Harry Diaz, and Suren Kulshreshtha, 1st ed. (Regina: University of Regina Press, 2010).

58 Woolway et al., "Lake Ice Will Be Less Safe for Recreation and Transportation Under Future Warming."

59 Clark et al., "Due North," 36.

60 NASA Global Climate Change, "Arctic Sea Ice Minimum," Climate Change: Vital Signs of the Planet, accessed July 16, 2023, https://climate.nasa.gov/vital-signs/arctic-sea-ice.

61 Bob Weber, "Climate Conference Hears Loss of Arctic Summer Sea Ice Now Inevitable by 2050," CBC News, November 2022, https://www.cbc.ca/news/canada/north/arctic-sea-ice-cop-27-1.6643783.

62 OARUSEPA, "Climate Change Indicators: Arctic Sea Ice," Reports and Assessments, July 2016, https://www.epa.gov/climate-indicators/climate-change-indicators-arctic-sea-ice.

63 Intergovernmental Panel on Climate Change, "AR6 Synthesis Longer Report," 40.

64 Intergovernmental Panel on Climate Change, 42.

65 Scott A. Kulp and Benjamin H. Strauss, "New Elevation Data Triple Estimates of Global Vulnerability to Sea-Level Rise and Coastal Flooding," *Nature Communications* 10, no. 1 (October 2019): 4844, doi:10.1038/s41467-019-12808-z. See also Tracey Wade, "Overview of Canadian Communities Exposed to Sea Level Rise," National Collaborating Centre for Environmental Health, August 2022, https://ncceh.ca/resources/evidence-reviews/overview-canadian-communities-exposed-sea-level-rise.

66 See, for example, City of Vancouver, "Climate Change and Sea Level Rise," 2023, https://vancouver.ca/green-vancouver/climate-change-and-sea-level-rise.aspx.

67 Sara Barron et al., "Delta-RAC Sea Level Rise Adaptation Visioning Study," 2012, 12, https://www.fraserbasin.bc.ca/_Library/CCAQ_BCRAC/bcrac_delta_visioning-policy_4d.pdf.

68 Barron et al., "Delta-RAC Sea Level Rise Adaptation Visioning Study."

69 Intergovernmental Panel on Climate Change, "AR6 Synthesis Longer Report."

70 Cheryl Santa Maria, "Fiona Is Atlantic Canada's Costliest Storm on Record, According to New Report," The Weather Network, October 2022, https://www.theweathernetwork.com/en/news/weather/severe/fiona-atlantic-canadas-costliest-storm-on-record-according-to-new, citing "Hurricane Fiona Causes $660 Million in Insured Damage," accessed May 15, 2023, http://www.ibc.ca/nl/resources/media-centre/media-releases/hurricane-fiona-causes-660-million-in-insured-damage.

71 Copernicus Climate Change Service, "Record-Breaking North Atlantic Ocean Temperatures Contribute to Extreme Marine Heatwaves | Copernicus," 2023, https://climate.copernicus.eu/record-breaking-north-atlantic-ocean-temperatures-contribute-extreme-marine-heatwaves; Hans Hersbach et al., "ERA5 Hourly Data

on Single Levels from 1940 to Present" (Copernicus Climate Change Service (C3S) Climate Data Store (CDS), 2023), doi:https://doi.org/10.24381/cds.adbb2d47.

72 The term "fire weather" has become ubiquitous, but is a title of a new book, John Vaillant, *Fire Weather: The Making of a Beast* (Knopf Canada, 2023), which set the term in my mind as I was writing this section.

73 Megan C. Kirchmeier-Young et al., "Attributing Extreme Fire Risk in Western Canada to Human Emissions," *Climatic Change* 144, no. 2 (September 2017): 365–379, doi:10.1007/s10584-017-2030-0.

74 John Vaillant, "No Wonder Alberta Is on Fire. We Made This Planet into a Volcano," *The Globe and Mail*, May 2023, https://www.theglobeandmail.com/opinion/article-no-wonder-alberta-is-on-fire-we-made-this-planet-into-a-volcano/.

75 Kristina A. Dahl et al., "Quantifying the Contribution of Major Carbon Producers to Increases in Vapor Pressure Deficit and Burned Area in Western US and Southwestern Canadian Forests," *Environmental Research Letters* 18, no. 6 (May 2023): 064011, doi:10.1088/1748-9326/acbce8.

76 Xianli Wang et al., "Projected Changes in Daily Fire Spread across Canada over the next Century," *Environmental Research Letters* 12, no. 2 (February 2017): 025005, doi:10.1088/1748-9326/aa5835.

77 James Hansen, Makiko Sato, and Reto Ruedy, "Perception of Climate Change," *Proceedings of the National Academy of Sciences* 109, no. 37 (September 2012), doi:10.1073/pnas.1205276109.

78 "Climate Change Sends Beetles Into Overdrive," accessed May 15, 2023, https://www.science.org/content/article/climate-change-sends-beetles-overdrive.

79 Natural Resources Canada, "Impacts" (Natural Resources Canada, July 2013), https://natural-resources.canada.ca/climate-change-adapting-impacts-and-reducing-emissions/climate-change-impacts-forests/impacts/13095.

80 Forest fire researcher Mike Flannigan, as quoted in Prairie Climate Centre, University of Winnipeg, "Forest Fires and Climate Change," Climate Atlas of Canada, 2022, https://climateatlas.ca/forest-fires-and-climate-change.

81 Brodie Thomas, "Council Is Told Calgary's 'future Is Smoky' as Climate Change Contributes to Wildfires," Calgary Herald, May 2023, https://calgaryherald.com/news/local-news/emergency-management-heat-events-threat.

82 Government of Canada, Department of Environment and Climate Change, "Air Quality," May 2012, https://www.canada.ca/en/environment-climate-change/services/environmental-indicators/air-quality.html; Matthew McClearn, "Wildfires Are Impairing Air Quality across Western Canada, Reversing Decades of Progress," *The Globe and Mail*, May 2023, https://www.theglobeandmail.com/canada/article-wildfires-are-impairing-air-quality-across-western-canada-reversing/.

83 United States Environmental Protection Agency, "Wildfires and Indoor Air Quality (IAQ)," December 2018, https://www.epa.gov/indoor-air-quality-iaq/wildfires-and-indoor-air-quality-iaq.

84 Jill Korsiak et al., "Long-Term Exposure to Wildfires and Cancer Incidence in Canada: A Population-Based Observational Cohort Study," *The Lancet Planetary Health* 6, no. 5 (May 2022): e400–e409, doi:10.1016/S2542-5196(22)00067-5; Rongbin Xu et al., "Wildfires, Global Climate Change, and Human Health," *New England Journal of Medicine* 383, no. 22 (November 2020): 2173–2181, doi:10.1056/NEJMsr2028985; "Critical Review of Health Impacts of Wildfire Smoke Exposure | Environmental Health Perspectives | Vol. 124, No. 9," accessed June 11, 2023, https://ehp.niehs.nih.gov/doi/10.1289/ehp.1409277.

85 "Smoke Sends US Northeast Solar Power Plunging by 50% as Wildfires Rage -
    Bloomberg," accessed June 11, 2023, https://www.bloomberg.com/news/articles/
    2023-06-08/smoke-sends-northeast-solar-power-plunging-by-50-as-wildfires-rage.

86 Intergovernmental Panel on Climate Change, "AR6 Synthesis Longer Report," 12.

87 Martha M. Vogel et al., "Concurrent 2018 Hot Extremes Across Northern
    Hemisphere Due to Human-Induced Climate Change," *Earth's Future* 7, no. 7
    (2019): 692–703, doi:10.1029/2019EF001189.

88 Quirin Schiermeier, "Climate Change Made Europe's Mega-Heatwave Five Times
    More Likely," *Nature* 571, no. 7764 (July 2019): 155–155, doi:10.1038/d41586-019-
    02071-z.

89 Sjoukje Y. Philip et al., "Rapid Attribution Analysis of the Extraordinary Heat
    Wave on the Pacific Coast of the US and Canada in June 2021," *Earth System
    Dynamics* 13, no. 4 (December 2022): 1689–1713, doi:10.5194/esd-13-1689-2022.

90 "Flood Vulnerability and Climate Change," *Canadian Climate Institute*
    (blog), accessed May 16, 2023, https://climateinstitute.ca/publications/flood-
    vulnerability-and-climate-change/.

91 Ryan Ness et al., "Under Water: The Costs of Climate Change for Canada's
    Infrastructure," Canadian Climate Institute, 2021, 12, https://climateinstitute.ca/
    reports/under-water/.

92 Environment Minister Catherine McKenna as quoted in Katie Dangerfield, "'100-
    Year Floods' Are Increasing in Canada Due to Climate Change, Officials Say — Is
    This True?," Global News, accessed May 15, 2023, https://globalnews.ca/news/
    5206116/100-year-floods-canada-increasing/.

93 Public Health Agency of Canada, "Climate Change, Floods and Your Health,"
    backgrounders;education and awareness, 2021, https://www.canada.ca/en/public-
    health/services/health-promotion/environmental-public-health-climate-change/
    climate-change-public-health-factsheets-floods.html.

94 Bernardo Teufel et al., "Investigation of the 2013 Alberta Flood from Weather
    and Climate Perspectives," *Climate Dynamics* 48, no. 9 (May 2017): 2881–2899,
    doi:10.1007/s00382-016-3239-8.

95 Kit Szeto et al., "The 2014 Extreme Flood on the Southeastern Canadian Prairies,"
    Bulletin of the American Meteorological Society 96, no. 12 (December 2015):
    S20–S24, doi:10.1175/BAMS-D-15-00110.1.

96 For a discussion of the challenges with measuring economic impacts of clean-ups,
    see Andrew Leach, "Oil Spills Boosting the Economy? That's Even Dumber than
    It Sounds," Macleans, May 2014, https://macleans.ca/economy/economicanalysis/
    saying-oil-spills-boost-the-economy-is-even-dumber-than-it-sounds-leach/.

97 Aiman Ghori, "Will Canada Benefit from Climate Change?," *Canadian Climate
    Institute* (blog), April 2021, https://climateinstitute.ca/will-canada-benefit-from-
    climate-change/.

98 See, for example, estimates for the United States versus Canada in Ricke et al.,
    "Country-Level Social Cost of Carbon." In all scenarios considered, the median
    country-level social cost of carbon emissions is substantially higher for the United
    States.

99 United States Department of Defense, "Climate Risk Analysis," 2021, 8, https://
    media.defense.gov/2021/Oct/21/2002877353/-1/-1/0/DOD-CLIMATE-RISK-
    ANALYSIS-FINAL.PDF. Selected parts removed by author for brevity.

100 Serge Olivier Kotchi et al., "Earth Observation-Informed Risk Maps of the Lyme
    Disease Vector Ixodes Scapularis in Central and Eastern Canada," *Remote Sensing*

13, no. 3 (January 2021): 524, doi:10.3390/rs13030524; NASA Earth Observatory, "Mapping the Spread of Lyme Disease," June 2021, https://earthobservatory.nasa. gov/images/148482/mapping-the-spread-of-lyme-disease.

101 "Crop Pests and Climate Change," *Climate Data Canada* (blog), accessed June 11, 2023, https://climatedata.ca/case-study/the-brown-marmorated-stink-bug-in-quebec/.

102 Carlos Yañez-Arenas et al., "Mapping Current and Future Potential Snakebite Risk in the New World," *Climatic Change* 134, no. 4 (February 2016): 697–711, doi:10.1007/s10584-015-1544-6.

103 Christy Climenhaga, "Glacier Melt Is Past the Tipping Point in the Canadian Rockies and That's a Big Problem," CBC, May 2022, https://www.cbc.ca/newsinteractives/ features/glacier-melt-is-past-the-tipping-point-in-the-canadian-rockies-and-thats-a-big-problem.

104 Jim Bronskill, "CSIS Warns Climate Change Threatens Canadian Security, Prosperity," *The Globe and Mail*, March 2023, https://www.theglobeandmail.com/ canada/article-csis-warns-climate-change-threatens-canadian-security-prosperity/.

105 Lorrie Goldstein, "Chattering Class Can't Handle Truth about Trudeau's Carbon Tax," Toronto Sun, June 2023, https://torontosun.com/opinion/columnists/ goldstein-chattering-class-cant-handle-truth-about-trudeaus-carbon-tax.

106 Tony Keller, "How Big Are Canada's Carbon Emissions? Compared to China, We're a Rounding Error," *The Globe and Mail*, May 2023, https://www.theglobeandmail. com/business/commentary/article-how-big-are-canadas-carbon-emissions-compared-to-china-were-a-rounding/.

107 Nasreddine Ammar et al., "A Distributional Analysis of the Clean Fuel Regulations" (Office of the Parliamentary Budget Officer, May 2023), 9, https://www.pbo-dpb. ca/en/publications/RP-2324-004-S--distributional-analysis-clean-fuel-regulations--analyse-distributive-reglement-combustibles-propres.

108 Government of Canada, *Canada's National Report on Climate Change: Actions to Meet Commitments under the United Nations Framework Convention on Climate Change* (Ottawa: Government of Canada, 1994); Government of Canada, "Canada's Eighth National Communication and Fifth Biennial Report on Climate Change," January 2023, https://unfccc.int/sites/default/files/resource/Canada%20 NC8%20BR5%20EN.pdf.

109 Emissions data in this paragraph all summarized from Hannah Ritchie, Max Roser, and Pablo Rosado, "CO_2 and Greenhouse Gas Emissions," Our World in Data, May 2020, https://ourworldindata.org/greenhouse-gas-emissions.

110 Andrew Coyne, "After This Season of Fire, the Conservatives Must Make Their Peace with Carbon Pricing," *The Globe and Mail*, June 2023, https://www. theglobeandmail.com/opinion/article-after-this-season-of-fire-the-tories-must-make-their-peace-with-carbon/.

111 Brendan Frank, "Why 1.6% Matters," *Canada's Ecofiscal Commission* (blog), May 2018, https://ecofiscal.ca/2018/05/23/why-1-6-matters/.

112 Simon Evans, "Analysis: Which Countries Are Historically Responsible for Climate Change?," Carbon Brief, October 2021, https://www.carbonbrief.org/ analysis-which-countries-are-historically-responsible-for-climate-change/.

113 Our World in Data, "Per Capita CO_2 Emissions," 2023, https://ourworldindata. org/grapher/co-emissions-per-capita.

114 Our World in Data, "Per Capita Consumption-Based CO_2 Emissions," 2023, https://ourworldindata.org/grapher/consumption-co2-per-capita.

115 Seth Klein, *A Good War: Mobilizing Canada for the Climate Emergency* (Toronto, Canada: ECW Press, 2020).

116 Employment and Social Development Canada, "Notes for an Address by The Right Honourable Stephen Harper, Prime Minister of Canada To the APEC Business Summit," September 2007, https://www.canada.ca/en/news/archive/2007/09/notes-address-right-honourable-stephen-harper-prime-minister-canada-apec-business-summit.html. Author's slight paraphrase and change of punctuation to convert from speaking notes to paragraph text.

117 George Hoberg, "How the Battles over Oil Sands Pipelines Have Transformed Climate Politics," preprint (Politics and International Relations, September 2019), doi:10.33774/apsa-2019-m4zgg.

118 Robert Tuttle, "BP's Oilsands Exit May Not Be the Last as Big Oil Revises Its Image," Bloomberg News via Financial Post, June 2022, https://financialpost.com/commodities/energy/oil-gas/bps-oil-sands-exit-may-not-be-the-last-as-big-oil-revises-image. Conoco-Phillips recently increased its oil sands holdings with the purchase of a 50% share in the Surmont project, after having sold other assets in 2017. See Mrinalika Roy and Rod Nickel, "ConocoPhillips to Buy Rest of Canada's Surmont Oil Site, Bumping Suncor," *Reuters*, May 2023, sec. Energy, https://www.reuters.com/business/energy/conocophillips-buys-remaining-stake-canadas-surmont-project-3-bln-2023-05-26/, with the 2017 asset sale discussed in Adam Samson and Ed Crooks, "ConocoPhillips Sells Oil Sands Assets for $13.3bn," *Financial Times*, March 2017, sec. US & Canadian companies, https://www.ft.com/content/38e9a030-14d8-11e7-b0c1-37e417ee6c76.

119 Canadian Energy Regulator, "C11610 Trans Mountain Pipeline ULC - Request for Confidential Treatment," April 2021, https://apps.cer-rec.gc.ca/REGDOCS/Item/View/4049977.

120 For a listing of commitments by country, see Ian W. H. Parry, Simon Black, and James Roaf, "Proposal for an International Carbon Price Floor Among Large Emitters," IMF, 2021, tbl. 1, https://www.imf.org/en/Publications/staff-climate-notes/Issues/2021/06/15/Proposal-for-an-International-Carbon-Price-Floor-Among-Large-Emitters-460468.

121 Ian Parry, Matt Davies, and Victor Mylonas, "Fiscal Policies for Paris Climate Strategies—from Principle to Practice," IMF, 2019, 27, https://www.imf.org/en/Publications/Policy-Papers/Issues/2019/05/01/Fiscal-Policies-for-Paris-Climate-Strategies-from-Principle-to-Practice-46826.

122 For a discussion of the target math at the time of the Kyoto protocol, see Jeffrey Simpson, Mark Jaccard, and Nic Rivers, *Hot Air: Meeting Canada's Climate Change Challenge* (McClelland & Stewart, 2011).

123 Parry, Black, and Roaf, "Proposal for an International Carbon Price Floor Among Large Emitters."

124 World Bank, "State and Trends of Carbon Pricing 2023," May 2023, 19–21, https://openknowledge.worldbank.org/handle/10986/39796. The policies referenced here are the "Clean Fuel Regulations," SOR/2022-140, Canada Gazette, Part 2, Volume 156, Number 14 § (2022), https://www.gazette.gc.ca/rp-pr/p2/2022/2022-07-06/html/sor-dors140-eng.html, and the "Regulations Amending the Reduction of Carbon Dioxide Emissions from Coal-Fired Generation of Electricity Regulations SOR/2018-263," Canada Gazette, Part II, Volume 152, Number 25 § (2018), https://gazette.gc.ca/rp-pr/p2/2018/2018-12-12/html/sor-dors263-eng.html.

125 Graham Thomson, "Climate Change Panel Tasked with Precarious Highwire Act,"

Edmonton Journal, accessed June 19, 2023, https://edmontonjournal.com/news/ politics/thomson-climate-change-panel-tasked-with-precarious-highwire-act.

126 Dale Smith, "Climate Target Bingo," April 2021, https://www.youtube.com/ watch?v=TrPI9Vzb2Sc.

127 Brian Murray and Nicholas Rivers, "British Columbia's Revenue-Neutral Carbon Tax: A Review of the Latest 'Grand Experiment' in Environmental Policy," *Energy Policy* 86 (November 2015): 674–683, doi:10.1016/j.enpol.2015.08.011.

128 Nathaniel O. Keohane, "Cap and Trade, Rehabilitated: Using Tradable Permits to Control U.S. Greenhouse Gases," *Review of Environmental Economics and Policy* 3, no. 1 (2009): 42–62, doi:10.1093/reep/ren021.

129 Emissions reductions achieved through reduced industrial output are implicitly valued at a lower amount because reducing output also reduces the allocation of free emissions credits in addition to reducing the carbon price payable.

130 Nasreddine Ammar et al., "A Distributional Analysis of Federal Carbon Pricing under A Healthy Environment and A Healthy Economy" (Office of the Parliamentary Budget Officer, March 2022), 10, https://www.pbo-dpb.ca/en/publications/ RP-2122-032-S--distributional-analysis-federal-carbon-pricing-under-healthy-environment-healthy-economy--une-analyse-distributive-tarification-federale-carbone-dans-cadre-plan-un-environnement-sain-une-eco.

131 Leah C. Stokes and Matto Mildenberger, "The Trouble with Carbon Pricing," Boston Review, 2020, https://www.bostonreview.net/articles/trouble-carbon-pricing/.

132 The federal carbon pricing backstop measures provincial policy stringency or *effort* by comparing the expected emissions reductions achieved by a suite of policies to what would be accomplished by carbon pricing. If the expected emissions reductions are not large enough, the federal carbon price will apply to the extent required to correct the shortfall in *effort*. Government of Canada, "Update to the Pan-Canadian Approach to Carbon Pollution Pricing 2023–2030," August 2021, https://www.canada.ca/en/environment-climate-change/services/ climate-change/pricing-pollution-how-it-will-work/carbon-pollution-pricing-federal-benchmark-information/federal-benchmark-2023-2030.html.

133 For a listing of countries where carbon pricing regimes are implemented, see World Bank, "State and Trends of Carbon Pricing 2023."

134 Government of Canada, "NC8-BR5."

135 For a summary of the expected emissions and energy impacts of the Inflation Reduction Act between now and 2035, see John Bistline et al., "Emissions and Energy Impacts of the Inflation Reduction Act," *Science* 380, no. 6652 (June 2023): 1324–1327, doi:10.1126/science.adg3781.

136 Clean Prosperity, "New Poll Shows Voters Still Expect a Credible Climate Plan," Clean Prosperity, July 2022, https://cleanprosperity.ca/new-poll-shows-voters-still-expect-a-credible-climate-plan/.

137 Modified from Frank, "Why 1.6% Matters." The full quote appears earlier in this Chapter.

138 International Energy Agency, "Greenhouse Gas Emissions from Energy," 2023, https://www.iea.org/data-and-statistics/data-tools/greenhouse-gas-emissions-from-energy-data-explorer.

139 Shannon Hall, "Exxon Knew about Climate Change Almost 40 Years Ago," Scientific American, 2015, https://www.scientificamerican.com/article/exxon-knew-about-climate-change-almost-40-years-ago/; Geoffrey Supran and Naomi

Oreskes, "Assessing ExxonMobil's Climate Change Communications (1977–2014)," *Environmental Research Letters* 12, no. 8 (August 2017): 084019, doi:10.1088/1748-9326/aa815f; Geoffrey Supran and Naomi Oreskes, "Rhetoric and Frame Analysis of ExxonMobil's Climate Change Communications," *One Earth* 4, no. 5 (May 2021): 696–719, doi:10.1016/j.oneear.2021.04.014; G. Supran, S. Rahmstorf, and N. Oreskes, "Assessing ExxonMobil's Global Warming Projections," *Science* 379, no. 6628 (January 2023): eabk0063, doi:10.1126/science.abk0063.

140 For an example, see Government of Canada, Department of Natural Resources, "Oil Sands: GHG Emissions" (Natural Resources Canada, June 2016), https://natural-resources.canada.ca/energy/publications/18731. Ezra Levant, *Ethical Oil: The Case for Canada's Oil Sands* (McClelland & Stewart, 2010), informed the federal government's oil sands agenda during the last Harper government. See, for example, Steven Chase, "Peter Kent's Green Agenda: Clean up Oil Sands' Dirty Reputation," *The Globe and Mail*, January 2011, https://www.theglobeandmail.com/news/politics/peter-kents-green-agenda-clean-up-oil-sands-dirty-reputation/article560974/.

141 Canadian Association of Petroleum Producers, "Made from Oil and Natural Gas," Context Magazine by CAPP, accessed November 22, 2022, https://context.capp.ca/infographics/2020/infographic_made-from-oil-and-natural-gas/.

142 Gurpreet Lail, President and CEO of the Petroleum Services Association of Canada, as quoted in Geoffrey Morgan, "'The World Still Needs Fossil Fuels:' Canada's Oilpatch Sees Future for the Industry despite 'Death Knell' Warning," *Financial Post*, August 2021, https://financialpost.com/commodities/the-world-still-needs-fossil-fuels-canadas-oilpatch-sees-future-for-the-industry-despite-death-knell-climate-warning.

143 CBC Radio, "The Current, Monday March 6, 2023 Episode Transcript," CBC, March 2023, https://www.cbc.ca/radio/thecurrent/monday-march-6-2023-episode-transcript-1.6770155.

144 ExxonMobil, "Outlook for Energy," ExxonMobil, October 2022, https://corporate.exxonmobil.com:443/what-we-do/energy-supply/outlook-for-energy.

145 "Cutting Oil and Gas Production Is Not Healthy, Says Shell Boss Wael Sawan," accessed May 17, 2023, https://www.thetimes.co.uk/article/cutting-oil-and-gas-production-is-not-healthy-says-shell-boss-wael-sawan-256m0ds7z.

146 bp, "Energy Outlook 2023," 2023, https://www.bp.com/content/dam/bp/business-sites/en/global/corporate/pdfs/energy-economics/energy-outlook/bp-energy-outlook-2023.pdf.

147 ExxonMobil, "Outlook for Energy", specifically the emissions graphic captured at perma.cc/8GBY-KB75.

148 Daniel Kahneman, *Thinking, Fast and Slow* (Doubleday Canada, 2011).

149 International Energy Agency, "WEO 2022," 330, Figure 7.1. Modifications to legend position made by author.

150 International Energy Agency, "WEO 2022."

151 International Energy Agency, 63–64.

152 Byers et al., "AR6 Scenarios Database."

153 Edward Byers et al., "AR6 Scenarios Database," April 2022, doi:10.5281/zenodo.5886912, which includes data on all simulations used to inform the Intergovernmental Panel on Climate Change, "AR6 Synthesis Longer Report" and related working group reports. Scenarios C1, C4, C6 and C7 are defined in Intergovernmental Panel on Climate Change, 2022: Summary for Policymakers [P.R. Shukla, J. Skea, A. Reisinger,

R. Slade, R. Fradera, M. Pathak, A. Al Khourdajie, M. Belkacemi, R. van Diemen, A. Hasija, G. Lisboa, S. Luz, J. Malley, D. McCollum, S. Some, P. Vyas, (eds.)]. In: Climate Change 2022: Mitigation of Climate Change. Contribution of Working Group III to the Sixth Assessment Report of the Intergovernmental Panel on Climate Change [P.R. Shukla, J. Skea, R. Slade, A. Al Khourdajie, R. van Diemen, D. McCollum, M. Pathak, S. Some, P. Vyas, R. Fradera, M. Belkacemi, A. Hasija, G. Lisboa, S. Luz, J. Malley, (eds.)]. Cambridge University Press, Cambridge, UK and New York, NY, USA. doi: 10.1017/9781009157926.001, 18-19.

154 International Energy Agency, "Oil 2023," 11.
155 Acknowledgement: This section is derived in part from an article published in *The Canadian Foreign Policy Journal*, November, 2022, copyright held by the Norman Paterson School of International Affairs, available online: https://www.tandfonline.com/doi/full/10.1080/11926422.2022.2120508.
156 International Energy Agency, "World Energy Investment 2023," 2023, 67, https://www.iea.org/reports/world-energy-investment-2023/overview-and-key-findings; International Energy Agency, "World Energy Investment 2020," 2020, 24, https://www.iea.org/reports/world-energy-investment-2023/overview-and-key-findings.
157 International Energy Agency, "World Energy Investment 2023," 62.
158 International Energy Agency, "WEO 2022," 357.
159 International Energy Agency, "World Energy Investment 2023," 64.
160 International Energy Agency, 74; International Energy Agency, "WEO 2022," 369. The IEA report estimates that global gas liquefaction capacity will rise to over 800 billion cubic metres per year by 2026.
161 International Energy Agency, "WEO 2022," 383.
162 International Energy Agency, 330, Figure 7.1. Modifications to legend position made by author.
163 Canadian Energy Regulator, "Canada's Energy Future 2023: Energy Supply and Demand Projections to 2050," 2023, https://www.cer-rec.gc.ca/en/data-analysis/canada-energy-future/2023/index.html; Canadian Energy Regulator, "Data From Canada's Energy Future 2023: Energy Supply and Demand Projections to 2050," Open Government Portal, accessed July 3, 2023, https://open.canada.ca/data/dataset/7643c948-d661-4d90-ab91-e9ac732fc737.
164 Government of Canada, Department of Environment and Climate Change, "2023 National Inventory Data," 2023, https://data-donnees.ec.gc.ca/data/substances/monitor/canada-s-official-greenhouse-gas-inventory/; Government of Canada, Department of Environment and Climate Change, "Canada's Greenhouse Gas and Air Pollutant Emissions Projections 2021," April 2022, https://publications.gc.ca/collections/collection_2022/eccc/En1-78-2021-eng.pdf.
165 Statistics Canada, "Canada's Natural Resource Wealth, 2021 (Preliminary Data)," The Daily, November 2022, https://www150.statcan.gc.ca/n1/daily-quotidien/221114/dq221114d-eng.htm.
166 Joseph Marchand, "Local Labor Market Impacts of Energy Boom-Bust-Boom in Western Canada," *Journal of Urban Economics* 71, no. 1 (January 2012): 165–174, doi:10.1016/j.jue.2011.06.001.
167 Amanda Stephenson, "Boomtown No More: How Alberta's Economy Has Changed, in Spite of Sky-High Oil Prices," Canadian Press, October 2022.
168 Statistics Canada, "Canada's Natural Resource Wealth, 2021 (Preliminary Data)." Specific data by resource in Statistics Canada. Table 38-10-0006-01: Value of selected natural resource reserves (x 1,000,000).

169 G. Kent Fellows, "Last Barrel Standing? Confronting the Myth of 'High-Cost' Canadian Oil Sands Production," CD Howe Institute Commentary (Toronto, ON, December 2022), https://papers.ssrn.com/abstract=4393421.

170 Fellows.

171 Aaron Cosbey, "The Bottom Line | Why Canada Is Unlikely to Sell the Last Barrel of Oil," *IISD Bottom Line* (blog), December 2022, https://www.iisd.org/articles/deep-dive/why-canada-unlikely-sell-last-barrel-oil.

172 Mohammad S. Masnadi et al., "Global Carbon Intensity of Crude Oil Production," *Science* 361, no. 6405 (August 2018): 851, doi:10.1126/science.aar6859; Mohammad S. Masnadi et al., "Carbon Implications of Marginal Oils from Market-Derived Demand Shocks," *Nature* 599, no. 7883 (November 2021): 80–84, doi:10.1038/s41586-021-03932-2.

173 Government of Canada, Department of Environment and Climate Change, "Greenhouse Gas Reporting Program (GHGRP) Facility Greenhouse Gas Data," 2023, https://open.canada.ca/data/en/dataset/a8ba14b7-7f23-462a-bdbb-83b0ef629823.

174 Calculations here rely on Alberta's industrial emissions policies, the "Technology Innovation and Emissions Reduction Regulation," Alta Reg 133-2019 § (2019), http://www.qp.alberta.ca/570.cfm?frm_isbn=9780779813278&search_by=link.

175 Dave Sawyer, "A Timbit with That Double-Double? Costs and Emission Reductions of Renewed Carbon Policy in Alberta," *IISD Policy Brief* (blog), 2014, https://www.iisd.org/system/files/publications/costs_emission_reductions_renewed_carbon_policy_alberta.pdf.

176 Government of Alberta, "Alberta Oil Sands Royalty Data," 2022, https://open.alberta.ca/opendata/alberta-oil-sands-royalty-data1.

177 Author's calculations using the model developed for Branko Bošković and Andrew Leach, "Leave It in the Ground? Oil Sands Development under Carbon Pricing," *Canadian Journal of Economics/Revue Canadienne d'économique* 53, no. 2 (2020): 526–562, prices from Sproule, "April 30, 2023 Escalated Forecast," Sproule, accessed July 15, 2023, https://sproule.com/wp-content/uploads/2023/05/2023-04-Escalated.xlsx, and project data from Government of Alberta, "Oil Sands Royalty Data". The exact break-even price will vary depending on a number of factors including Canadian heavy crude oil discounts to global benchmark light oil prices, the Canadian-US dollar exchange rate, and the price of natural gas.

178 Roy and Nickel, "ConocoPhillips to Buy Rest of Canada's Surmont Oil Site, Bumping Suncor."

179 Government of Canada, "Options to Cap and Cut Oil and Gas Sector Greenhouse Gas Emissions to Achieve 2030 Goals and Net-Zero by 2050," July 2022, https://www.canada.ca/en/services/environment/weather/climatechange/climate-plan/oil-gas-emissions-cap/options-discussion-paper.html; Government of Canada, "Clean Fuel Regulations SOR/2022-140," Canada Gazette, Part II, Vol 156, N 14 § (2022), https://www.gazette.gc.ca/rp-pr/p2/2022/2022-07-06/html/sor-dors140-eng.html.

180 Tuttle, "BP's Oilsands Exit May Not Be the Last as Big Oil Revises Its Image". Conoco-Phillips recently increased its oil sands holdings with the purchase of a 50% share in the Surmont project, after having sold other assets in 2017. See Roy and Nickel, "ConocoPhillips to Buy Rest of Canada's Surmont Oil Site, Bumping Suncor", with the 2017 asset sale discussed in Samson and Crooks, "ConocoPhillips Sells Oil Sands Assets for $13.3bn."

181 Reuters, "Shell Sells Oil Sands Assets as Boss Warns on Clean Energy Challenge," *The*

Guardian, March 2017, sec. World news, https://www.theguardian.com/world/2017/mar/10/shell-sells-canadian-oil-sands-as-boss-warns-of-losing-public-support.

182 Swiss Re, "We Are Aligning Our Oil and Gas Business to Our Net-Zero Commitment," March 2022, https://www.swissre.com/sustainability/stories/aligning-oil-gas-business.html.

183 "New Oil & Gas Investment / Underwriting Guidelines | Munich Re," accessed May 18, 2023, https://www.munichre.com/en/company/media-relations/statements/2022/new-oil-and-gas-investment-underwriting-guidelines.html.

184 United Nations Environment Programme, "Global Insurance and Reinsurance Leaders Establish Groundbreaking Net-Zero Alliance," July 2021, https://www.unepfi.org/industries/insurance/global-insurance-and-reinsurance-leaders-establish-alliance-to-accelerate-transition-to-net-zero-emissions-economy/.

185 Canadian Energy Regulator, "C11610 Trans Mountain Pipeline ULC - Request for Confidential Treatment."

186 Jeff Lewis, "Insurers and Banks That Have Shunned Canada's Oil Sands," Insurance Journal, October 2020, https://www.insurancejournal.com/news/international/2020/10/15/586602.htm. Note that the Alberta Government has developed legislation which may allow oil sands companies to self-insure through captive insurance companies (*Captive Insurance Companies Act*, SA 2021, c C-2.4).

187 Oil Sands Pathways to Net Zero Initiative, "Pathways Plan to Achieve Net-Zero Emissions," Oil Sands Pathways to Net Zero, November 2021, https://www.oilsandspathways.ca/the-pathways-vision/.

188 Kyra Bell-Pasht and 2022, "What Is the Path to Net-Zero Emissions for Oilsands Producers?," Policy Options, February 2022, https://policyoptions.irpp.org/magazines/february-2022/what-is-the-path-to-net-zero-emissions-for-oilsands-producers/; Gabriel Friedman, "'Dangerous Distraction' or Silver Bullet? Opinion Divided on Government's Role in Carbon Capture Investments," Financial Post, July 2021, https://financialpost.com/commodities/energy/oil-gas/is-carbon-capture-the-most-cost-effective-way-to-decarbonize-canadas-oilpatch.

189 Government of Canada, *Canada's First National Report on Climate Change*.

190 Alberta Carbon Capture and Storage and Development Council, "Accelerating Carbon Capture and Storage Implementation in Alberta" (Government of Alberta, March 2009).

191 Government of Canada, "Turning the Corner: Taking Action to Fight Climate Change," 2008, https://publications.gc.ca/site/archivee-archived.html?url=http://publications.gc.ca/collections/collection_2009/ec/En88-2-2008E.pdf.

192 Government of Alberta, "Carbon Capture, Utilization and Storage—Development and Innovation," 2023, https://www.alberta.ca/carbon-capture-utilization-and-storage-development-and-innovation.aspx.

193 CBC Radio, "The Current, Monday March 6, 2023 Episode Transcript."

194 Rod Nickel, "Oil Companies Ask Canada to Pay for 75% of Carbon Capture Facilities," October 2021, https://www.reuters.com/world/americas/exclusive-oil-companies-ask-canada-pay-75-carbon-capture-facilities-2021-10-07/.

195 West of Centre, "'Walking the Walk': A Major Oil CEO on Emissions Targets, COP26 and New Ministers," CBC Podcasts, October 2021, https://www.cbc.ca/listen/cbc-podcasts/407-west-of-centre/episode/15875444-walking-the-walk-a-major-oil-ceo-on-emissions-targets-cop26-and-new-ministers. The explanation of the $3 trillion value in the interview is not as clear as I would like, but similar claims of $3–4 trillion GDP contributions are common in the Canadian oil and

gas industry so I have paraphrased it as such here. See, for example "Alberta's Oil Sands Expected Contribute \$4 Trillion to Canada's GDP over the next 25 Years," Oil Sands Magazine, November 2014, https://www.oilsandsmagazine.com/news/2014/11/12/alberta-oil-sands-expected-contribute-4-trillion-to-canadian-gdp-over-next-25-year.

196 Among the four major oil sands companies (Suncor, Canadian Natural, Imperial Oil, and Cenovus), only Cenovus does not have oil sands mining assets.

197 Alberta Energy Regulator, "Manual 024: Guide to the Mine Financial Security Program," AER, 2022, v, https://static.aer.ca/prd/documents/manuals/Manual024.pdf.

198 Alberta Energy Regulator, 22.

199 Alberta Energy Regulator, 10–11.

200 Government of Alberta, "Financial Deposits for Reclamation Work under Review," May 2021, https://www.alberta.ca/news.aspx.

201 Alberta Energy Regulator, "Manual 024: Guide to the Mine Financial Security Program," iv.

202 Bjorn Lomborg, "Ill-Advised 'net-Zero' Emissions Policies Are Netting Worldwide Pain," National Post, October 2022, https://nationalpost.com/opinion/bjorn-lomborg-ill-advised-net-zero-emissions-policies-are-netting-worldwide-pain.

203 "Sun, Wind and Drain," The Economist, accessed May 29, 2023, https://www.economist.com/finance-and-economics/2014/07/29/sun-wind-and-drain.

204 Lazard, "Levelized Cost of Energy Analysis (v 16.0)," 2023, https://www.lazard.com/media/451905/lazards-levelized-cost-of-energy-version-150-vf.pdf. The 90% drop compares 2009 costs to 2021 costs. Recent inflation and supply-chain issues have led to increased costs in 2023 per the preliminary Lazard estimates.

205 Leah C. Stokes, "The Politics of Renewable Energy Policies: The Case of Feed-in Tariffs in Ontario, Canada," Energy Policy 56 (May 2013): 490–500, doi:10.1016/j.enpol.2013.01.009.

206 Clean Energy Canada, "A Renewables Powerhouse," Clean Energy Canada, 2023, https://cleanenergycanada.org/report/a-renewables-powerhouse/.

207 Rupert Way et al., "Empirically Grounded Technology Forecasts and the Energy Transition," Joule 6, no. 9 (September 2022): 2057–2082, doi:10.1016/j.joule.2022.08.009.

208 Auke Hoekstra, "Photovoltaic Growth: Reality versus Projections of the International Energy Agency – with 2018 Update (by Auke Hoekstra)," Steinbuch (blog), June 2017, https://maartensteinbuch.com/2017/06/12/photovoltaic-growth-reality-versus-projections-of-the-international-energy-agency/; Way et al., "Empirically Grounded Technology Forecasts and the Energy Transition."

209 Alberta Electric System Operator, "2021 Long Term Outlook," AESO, 2021, https://www.aeso.ca/assets/Uploads/grid/lto/2021-Long-term-Outlook.pdf.

210 Alberta Electric System Operator, "Reliability Requirements Roadmap," 2023, https://www.aeso.ca/assets/Uploads/future-of-electricity/AESO-2023-Reliability-Requirements-Roadmap.pdf.

211 International Renewable Energy Agency, "Renewable Capacity Statistics 2023," 2023, https://www.irena.org/Publications/2023/Mar/Renewable-capacity-statistics-2023; International Energy Agency, "Renewable Energy Market Update," June 2023, https://www.iea.org/reports/renewable-energy-market-update-june-2023.

212 US Department of Energy, "Quarterly Solar Industry Update," Energy.gov, May 2023, https://www.energy.gov/eere/solar/quarterly-solar-industry-update.

213 International Renewable Energy Agency, "Renewable Capacity Statistics 2023."
214 Sara Hastings-Simon et al., "Alberta's Renewable Electricity Program: Design, Results, and Lessons Learned," *Energy Policy* 171 (2022): 113266, doi:https://doi.org/10.1016/j.enpol.2022.113266.
215 Alberta Electric System Operator, "2019 Long-Term Outlook," AESO, 2019, https://www.aeso.ca/grid/grid-planning/forecasting/2019-long-term-outlook/.
216 Alberta Electric System Operator, "2023 Long Term Outlook Engagement Session Presentation," June 2023, https://perma.cc/425K-PGHU.
217 Alberta Electric System Operator, "2021 LTO."
218 Alberta Electric System Operator, "2023 Long Term Outlook Engagement Session Presentation."
219 Alberta Electric System Operator, "2016 Long-Term Outlook," AESO, 2017, http://www.aeso.ca/downloads/AESO_2016_Long-term_Outlook_WEB.pdf.
220 Lazard, "Levelized Cost of Energy Analysis (v 16.0)."
221 The plan is detailed in Government of Canada, "A Healthy Environment and a Healthy Economy," 2020, https://www.canada.ca/en/services/environment/weather/climatechange/climate-plan/climate-plan-overview/healthy-environment-healthy-economy.html, while the regulatory proposal is discussed in Government of Canada, "A Clean Electricity Standard in Support of a Net-Zero Electricity Sector: Discussion Paper," March 2022, https://www.canada.ca/en/environment-climate-change/services/canadian-environmental-protection-act-registry/achieving-net-zero-emissions-electricity-generation-discussion-paper.html.
222 Canada and US data as well as global averages from Małgorzata Wiatros-Motyka, "Global Electricity Review 2023," Ember Climate, April 2023, https://ember-climate.org/insights/research/global-electricity-review-2023/.
223 Canadian Energy Regulator, "Market Snapshot: Canada – 2nd in the World for Hydroelectric Production," July 2022, https://www.cer-rec.gc.ca/en/data-analysis/energy-markets/market-snapshots/2016/market-snapshot-canada-2nd-in-world-hydroelectric-production.html; Government of Canada, Department of Environment and Climate Change, "2023 NIR Data."
224 Government of Canada, Department of Natural Resources, "Canada's Wind TRM," June 2010, https://natural-resources.canada.ca/energy/renewable-electricity/wind/7323; National Renewable Energy Laboratory, "Wind Resource Maps and Data," accessed June 4, 2023, https://www.nrel.gov/gis/wind-resource-maps.html.
225 Manajit Sengupta et al., "The National Solar Radiation Data Base (NSRDB)," *Renewable and Sustainable Energy Reviews* 89 (June 2018): 51–60, doi:10.1016/j.rser.2018.03.003.
226 Government of Canada, Department of Environment and Climate Change, "2023 NIR Data."
227 Blake Shaffer, "Technical Pathways to Aligning Canadian Electricity Systems with Net-Zero Goals," Canadian Institute for Climate Choices, 2021, 8, https://climateinstitute.ca/wp-content/uploads/2021/09/CICC-Technical-pathways-to-aligning-Canadian-electricity-systems-with-net-zero-goals-by-Blake-Shaffer-FINAL-1.pdf.
228 Lazard, "Levelized Cost of Energy Analysis (v 16.0)."
229 Jacqueline A. Dowling et al., "Role of Long-Duration Energy Storage in Variable Renewable Electricity Systems," *Joule* 4, no. 9 (September 2020): 1907–1928, doi:10.1016/j.joule.2020.07.007.
230 For example, in California, current rates offered by Southern California Edison

have weekday peak charges of 65-74 US cents per kilowatt hour, and off-peak charges or 28-37 US cents per kilowatt hour, so shifting demand away from peak periods or using stored electricity in those periods would save more than 30 US cents per kilowatt hour relative to off-peak prices: Southern California Edison, "Time-Of-Use Residential Rate Plans," accessed June 7, 2023, https://www.sce.com/residential/rates/Time-Of-Use-Residential-Rate-Plans.

231 "The Case for an Environmentalism That Builds," *The Economist*, accessed May 13, 2023, https://www.economist.com/leaders/2023/04/05/the-case-for-an-environmentalism-that-builds.

232 Alberta's interconnections cannot be operated simultaneously at full capacity, but the details add too much complexity for this discussion.

233 Brett Dolter and Nicholas Rivers, "The Cost of Decarbonizing the Canadian Electricity System," *Energy Policy* 113 (February 2018): 135–148, doi:10.1016/j.enpol.2017.10.040.

234 Canadian Energy Regulator, "Towards Net-Zero: Electricity Scenarios," November 2022, https://www.cer-rec.gc.ca/en/data-analysis/canada-energy-future/2021/towards-net-zero.html.

235 See Andrew Pascale and Jesse D Jenkins, "Annex F: Integrated Transmission Line Mapping and Costing," in *Princeton's Net-Zero America Study*, 2021.

236 Paul Denholm et al., "Examining Supply-Side Options to Achieve 100% Clean Electricity by 2035," September 2022, doi:10.2172/1885591. See also Nadja Popovich and Brad Plumer, "Why the U.S. Electric Grid Isn't Ready for the Energy Transition," *The New York Times*, June 2023, sec. Climate, https://www.nytimes.com/interactive/2023/06/12/climate/us-electric-grid-energy-transition.html.

237 "What Is the Dreaded Dunkelflaute?," accessed May 29, 2023, https://qz.com/can-europe-survive-the-dreaded-dunkelflaute-1849886529.

238 Government of Ontario, "Powering Ontario's Growth: Ontario's Plan for a Clean Energy Future," 2023, http://www.ontario.ca/page/powering-ontarios-growth.

239 Canadian Energy Regulator, "Energy Futures 2023," 72.

240 International Energy Agency, "WEO 2022," 128. See also David Brown, "The Nuclear Option," Wood Mackenzie, May 2023, https://www.woodmac.com/horizons/making-new-nuclear-power-viable-in-the-energy-transition/, for similar results.

241 Canadian Energy Regulator, "Energy Futures 2023," 70.

242 Lazard, "Levelized Cost of Energy Analysis (v 16.0)," 31.

243 Energy Information Administration, "Annual Energy Outlook 2023," accessed October 16, 2022, https://www.eia.gov/outlooks/aeo/.

244 Brown, "The Nuclear Option."

245 J. D. Jenkins et al., "The Benefits of Nuclear Flexibility in Power System Operations with Renewable Energy," *Applied Energy* 222 (July 2018): 872–884, doi:10.1016/j.apenergy.2018.03.002.

246 Shaffer, "Technical Pathways to Aligning Canadian Electricity Systems with Net-Zero Goals", citing Dowling et al., "Role of Long-Duration Energy Storage in Variable Renewable Electricity Systems".

247 US Department of Energy, "Quarterly Solar Industry Update."

248 Capacity factors will be much lower for solar than for nuclear power, so this should not be interpreted as a claim that one year's worth of new solar installations will generate as much electricity as the global nuclear fleet. A reasonable expectation would be closer to one quarter as much generation, which is still substantial.

249 Dolter and Rivers, "The Cost of Decarbonizing the Canadian Electricity System."

250 "I Was Wrong About the Limits of Solar. PV Is Becoming Dirt Cheap," accessed May 14, 2023, https://www.greentechmedia.com/articles/read/i-was-wrong-about-the-economic-limitations-of-solar-power.

251 CBC News, "Hanna, Alta, Could Be Hit Hard by Coal Phase-Out," January 2017, https://www.cbc.ca/news/canada/calgary/hanna-alberta-coal-mine-shut-down-1.3936178.

252 "ATCO Plans to Convert Alberta Coal Plants to Gas by 2020," accessed May 9, 2023, https://www.spglobal.com/marketintelligence/en/news-insights/trending/B6xr0rf4CA4MWyAEeAuUlg2.

253 Government of Canada, Department of Natural Resources, "Briefing Package for the Minister of Natural Resources to Appear before the House of Commons Standing Committee on Natural Resources (RNNR) for Its Study on Creating a Fair and Equitable Canadian Energy Transformation," 2022, https://open.canada.ca/data/en/dataset/24ae60ef-359d-4c67-aa31-a71e5e7aa88d/resource/7900aa7c-401f-42ed-bb23-64d4b599cf86.

254 Government of Canada, Department of Natural Resources.

255 Government of Canada, Department of Natural Resources, "Energy Fact Book 2022-23," 2023, https://publications.gc.ca/collections/collection_2022/rncan-nrcan/M136-1-2022-eng.pdf.

256 Government of Canada, Department of Natural Resources.

257 Beata Caranci and Francis Fong, "Don't Let History Repeat: Canada's Energy Sector Transition and the Potential Impact on Workers," TD Economics, April 2021, https://economics.td.com/esg-energy-sector.

258 Jim Stanford, "Employment Transitions and the Phase-Out of Fossil Fuels" (Centre for Future Work, January 2021), 19.

259 "The New Reality," Clean Energy Canada (blog), 2021, https://cleanenergycanada.org/report/the-new-reality/.

260 David A. Green and Benjamin M. Sand, "Has the Canadian Labour Market Polarized?," The Canadian Journal of Economics / Revue Canadienne d'Economique 48, no. 2 (2015): 612–646; Nicole M. Fortin and Thomas Lemieux, "Changes in Wage Inequality in Canada: An Interprovincial Perspective," The Canadian Journal of Economics / Revue Canadienne d'Economique 48, no. 2 (2015): 682–713; Marchand, "Local Labor Market Impacts of Energy Boom-Bust-Boom in Western Canada."

261 Rachel Samson, "Just Transition or Smart Transition?," Policy Options, March 2023, https://policyoptions.irpp.org/magazines/march-2023/just-smart-transition/.

262 Stanford, "Employment Transitions and the Phase-Out of Fossil Fuels," 51.

263 Stanford, 51.

264 Office of the Auditor General, "Department of Fisheries and Oceans Northern Cod Adjustment and Recovery Program," in Report of the Auditor General to the House of Commons for 1993, 1993.

265 Heritage Newfoundland and Labrador, "Economic Impacts of the Cod Moratorium," 2008, https://www.heritage.nf.ca/articles/economy/moratorium-impacts.php.

266 Matt Maiorana, "Managed Decline: A Just Clean Energy Transition and Lessons from Canada's Cod Fishing Industry," Oil Change International (blog), September 2016, https://priceofoil.org/2016/09/12/managed-decline-a-just-clean-energy-transition-and-lessons-from-canadas-cod-fishing-industry/.

267 Statistics Canada. Table 25-10-0065-01 Oil and gas extraction revenues, expenses and balance sheet.

268 For example, Alberta resource revenue spiked to a record $25 billion in 2022-23.

269 Office of the Auditor General, "Department of Fisheries and Oceans Northern Cod Adjustment and Recovery Program," sec. 15.13.

270 "The Newfoundland Fishery - a Descriptive Analysis," accessed April 28, 2023, http://www.ucs.mun.ca/~noelroy/NfFishery.text.html. See also Statistics Canada. Table 14-10-0245-01 Average weekly earnings (SEPH), by type of employee for selected industries classified using the North American Industry Classification System (NAICS). DOI: https://doi.org/10.25318/1410024501-eng.

271 See Statistics Canada. Table 14-10-0245-01 Average weekly earnings (SEPH), by type of employee for selected industries classified using the North American Industry Classification System (NAICS). DOI: https://doi.org/10.25318/1410024501-eng

272 Office of the Auditor General, "Department of Fisheries and Oceans Northern Cod Adjustment and Recovery Program," sec. 15.9.

273 Heritage Newfoundland and Labrador, "Economic Impacts of the Cod Moratorium."

274 Emily Grubert, "What Happens to Gas Stations When the World Goes Electric?," TED Talk, accessed June 15, 2023, https://www.ted.com/talks/emily_grubert_what_happens_to_gas_stations_when_the_world_goes_electric/c. Canadian labour economist Jim Stanford makes a similar argument in his paper, "Employment Transitions and the Phase-Out of Fossil Fuels."

275 Helmuth Graf von Moltke, *Moltkes militärische Werke: Die Thätigkeit als Chef des Generalstabes der Armee im Frieden. Th. 1. Taktische Aufgaben aus den Jahren 1858 bis 1882. Th. 2. Taktisch-strategische Aufsätze aus den Jahren 1857 bis 1871. 2 v* (E. S. Mittler, 1900).

276 For example, the Genesee mine and generating station, combined, employed about 360 people for operations, Battle River approximately 280, and Sheerness approximately 230 people. (Government of Canada, "NPRI Data," accessed April 29, 2023, https://open.canada.ca/data/en/dataset/40e01423-7728-429c-ac9d-2954385ccdfb/resource/ed5b6b49-2e63-48e6-9418-adafa86ec6f5.)

277 All of the submissions to the Alberta advisory panel are publicly available. See "Climate Change Advisory Panel," Government of Alberta, 2015, https://www.alberta.ca/climate-leadership-discussion.aspx.

278 Government of Alberta, "REVISED: Alberta Announces Coal Transition Action," November 2016, https://www.alberta.ca/news.aspx.

279 "Getting It Right: A Just Transition for Alberta's Coal Workers," July 2017, https://web.archive.org/web/20170716190922/https://d3n8a8pro7vhmx.cloudfront.net/afl/pages/3043/attachments/original/1488233038/getting_it_right_lowres_feb27.pdf?1488233038; Government of Alberta and Advisory Panel on Coal Communities, "Supporting Workers and Communities: Recommendations to the Government of Alberta," accessed April 29, 2023, https://www.alberta.ca/assets/documents/advisory-panel-coal-communities-recommendations.pdf.

280 Government of Canada, Department of Environment and Climate Change, "Final Report by the Task Force on Just Transition for Canadian Coal Power Workers and Communities: A Just and Fair Transition for Canadian Coal Power Workers and Communities," 2018, 11, https://publications.gc.ca/site/eng/9.867000/publication.html.

281 Office of the Auditor General of Canada Government of Canada, "Report 1—Just

Transition to a Low-Carbon Economy," April 2022, 17, https://www.oag-bvg.gc.ca/internet/English/parl_cesd_202204_01_e_44021.html.

282 CBC News, "Coal Town Hopes Alberta Government Grants Will Help It Bounce Back | CBC News," CBC, August 2017, https://www.cbc.ca/news/canada/calgary/hanna-coal-alberta-grants-diversification-1.4266241.

283 The quotes which follow this one are all taken from McGowan, Gil, Testimony to the House of Commons Standing Committee on Natural Resources, number 18, 1st session 44th Parliament, Wednesday, April 27, 2022.

284 Economist Jim Stanford writes that "most of the coming decline in fossil fuel employment can be absorbed through normal attrition and retirement (…) if those normal departures are not replaced with new hires." Such a plan was implemented by the state-run Ruhrkohle consortium in Germany and is mentioned at two points later in this chapter, but there is no mechanism for governments to force oil and gas employers to hire workers displaced from other sites. See Stanford, "Employment Transitions and the Phase-Out of Fossil Fuels," 86.

285 McGowan, Gil, Testimony to the House of Commons Standing Committee on Natural Resources, number 18, 1st session 44th Parliament, Wednesday, April 27, 2022.

286 Stanford, "Employment Transitions and the Phase-Out of Fossil Fuels," 90.

287 On the broader topic of social safety net renewal, see Christopher McCabe et al., "Renewing the Social Contract: Economic Recovery in Canada from COVID-19," The Royal Society of Canada, 2020, https://rsc-src.ca/en/research-and-reports/covid-19-policy-briefing/economic-recovery/renewing-social-contract-economic.

288 Canada has allowed some equivalency agreements which may see coal units produce beyond 2030, but with carbon pricing in place it is plausible that we would see all coal units phased out before 2030 in Canada as proposed in the initial regulations.

289 Andrea Furnaro et al., "German Just Transition: A Review of Public Policies to Assist German Coal Communities in Transition" (Resources for the Future, 2021), 7, https://media.rff.org/documents/21-13-Nov-22.pdf.

290 Furnaro et al., 7.

291 "German Coal Imports Rose 8% in 2022, Russia Sold 37% Less - Data | Reuters," accessed May 10, 2023, https://www.reuters.com/markets/commodities/german-coal-imports-rose-8-2022-russia-sold-37-less-data-2023-02-27/.

292 Furnaro et al., "RFF German Coal," 42.

293 Furnaro et al., 42.

294 Furnaro et al., 42–62.

295 "Coal Mining Deal Struck," Deutsche Welle, February 2007, https://www.dw.com/en/final-deal-struck-to-end-german-subsidized-deep-coalmining/a-2341256.

296 Mark S. Winfield and Abdeali Saherwala, "Phasing Out Coal-Fired Electricity in Ontario," in Policy Success in Canada: Cases, Lessons, Challenges, ed. Evert Lindquist et al. (Oxford University Press, 2022), doi:10.1093/oso/9780192897046.003.0019; Mark Winfield, Blue-Green Province: The Environment and the Political Economy of Ontario (UBC Press, 2012).

297 Winfield and Saherwala, "Phasing Out Coal-Fired Electricity in Ontario."

298 Ontario Power Authority, "Power System Planning: Supply Mix Summary," 2005, http://www.powerauthority.on.ca/Report_Static/1139.htm, via Winfield and Saherwala, "Phasing Out Coal-Fired Electricity in Ontario."

299 Environment and Climate Change Canada, "Final Report by the Task Force on Just

Transition for Canadian Coal Power Workers and Communities," March 2022, 5, https://www.canada.ca/en/environment-climate-change/services/climate-change/task-force-just-transition/final-report.html.

300 Melissa Harris, Marisa Beck, and Ivetta Gerasimchuk, "The End of Coal: Ontario's Coal Phase-Out," IISD, 2015, https://www.iisd.org/system/files/publications/end-of-coal-ontario-coal-phase-out.pdf.

301 "How to Ditch Fossil Fuels without Leaving Workers Behind," *What on Earth* (CBC, April 2023), https://www.cbc.ca/listen/live-radio/1-429-what-on-earth/clip/15981519-how-ditch-fossil-fuels-without-leaving-workers-behind?share=true. Full disclosure: I spoke with producer Molly Segal in advance of the production of this episode.

302 Canadian Energy Regulator, "Canada's Energy Future 2021," 2021, https://www.cer-rec.gc.ca/en/data-analysis/canada-energy-future/2021/canada-energy-futures-2021.pdf.

303 "Managing a Just Transition in Denmark," *Canadian Climate Institute* (blog), accessed May 11, 2023, https://climateinstitute.ca/publications/managing-a-just-transition-in-denmark/.

304 Stanford, "Employment Transitions and the Phase-Out of Fossil Fuels."

305 Emily Grubert and Sara Hastings-Simon, "Designing the Mid-Transition: A Review of Medium-Term Challenges for Coordinated Decarbonization in the United States," *WIREs Climate Change* 13, no. 3 (2022): e768, doi:10.1002/wcc.768. See also Grubert, "Emily Grubert", from which I quoted earlier in this chapter.

306 Steven Chase and Les Perreaux, "Ottawa Does U-Turn on Asbestos Mining," Globe and Mail, September 2012, https://www.theglobeandmail.com/news/politics/ottawa-does-u-turn-on-asbestos-mining/article4545704/; Bill Chappell, "Canada Stops Its Defense Of Asbestos, As Quebec's Mines Close For Good," *NPR*, September 2012, sec. International, https://www.npr.org/sections/thetwo-way/2012/09/17/161298741/canada-stops-its-defense-of-asbestos-quebecs-mines-shut-down.

307 In the context of the recovery from COVID-19, McCabe et al., "Renewing the Social Contract: Economic Recovery in Canada from COVID-19" looks at many tools which are of general application for dealing with worker transitions and widespread displacement.

308 In Nigel Bankes and Andrew Leach, "Preparing for a Midlife Crisis: Section 92A at 40," forthcoming, Alberta Law Review, 2023, doi:10.2139/ssrn.4262958, we look at the history of judicial interpretation of provinicial legislative jursidiction over resources as defined in Section 92A.

309 Ammar et al., "A Distributional Analysis of Federal Carbon Pricing under A Healthy Environment and A Healthy Economy."

310 Mark Z. Jacobson, *No Miracles Needed: How Today's Technology Can Save Our Climate and Clean Our Air* (Cambridge: Cambridge University Press, 2023), 70–71, doi:10.1017/9781009249553.

311 "Comparative Life-Cycle Greenhouse Gas Emissions of a Mid-Size BEV and ICE Vehicle – Charts – Data & Statistics," accessed June 13, 2023, https://www.iea.org/data-and-statistics/charts/comparative-life-cycle-greenhouse-gas-emissions-of-a-mid-size-bev-and-ice-vehicle.

312 Zeke Hausfather, "Factcheck: How Electric Vehicles Help to Tackle Climate Change," Carbon Brief, May 2019, https://www.carbonbrief.org/factcheck-how-electric-vehicles-help-to-tackle-climate-change/.

313 Thanks to Hannah Ritchie of Our World in Data for a tweet in response to Rowan

Atkinson (the actor who played Mr. Bean) that led me to include this paragraph and from which I drew some of the sources for this brief section: June 2023, https://twitter.com/_HannahRitchie/status/1665448694936657922.

314 See, for example, analysis of the emissions impacts of the Trans Mountain Pipeline expansion: Government of Canada, Department of Environment and Climate Change, "Trans Mountain Pipeline ULC - Trans Mountain Expansion Project: Review of Related Upstream Greenhouse Gas Emissions Estimates," November 2016, https://perma.cc/T7YF-RP2E.

315 Canadian Securities Administrators, "51-107 - Climate-Related Disclosure Update and CSA Notice and Request for Comment Proposed National Instrument 51-107 Disclosure of Climate-Related Matters," 2021, https://perma.cc/UE6V-8GPB.